ASTROLOGY ESSENTIALS

ASTROLOGY
Essentials

Eleanor Haspel-Portner, Ph.D.

Astrology Essentials
Eleanor Haspel-Portner, Ph.D.

Copyright © 2017 Noble Sciences, LLC.

All rights reserved.
This book or any portion thereof may not be reproduced or used in any manner whatsoever without the express written permission of the publisher except for the use of brief quotations in a book review.

ISBN: 978-1-93105313-6

Other titles by Eleanor Haspel-Portner
Cosmic Secrets Revealed
Patterns of Orientation
*Marriage in Trouble: A Time of Decisio*n

Author's websites
www.beyondhumandesign.com
www.noblesciences.com
www.consciouslifechoices.com
www.DrEleanorHaspel-Portner.com
www.moptwo.com/DrEleanor

Illustrations created by Eleanor Haspel-Portner, Ph.D

Book Design by Michelle M. White
www.mmwgraphicdesign.com

Dedicated with deep love and gratitude to Katherine de Jersey, a truly great astrologer and inspiration.

This Book provides basic foundational information for understanding and working with Noble Sciences Tools. It is "what you need to know". However, many years of study are required to master the complexity and breadth of all Noble Sciences information. A Basic Key to the Color Codes used in all Noble Sciences Documents is included at the end of this book.

Contents

Illustration List
1

Foreword
5

PART I
Noble Sciences Sacred Synthesis
Components
7

PART II
Astrology for Noble Sciences
What You Need To Know
19

PART III
Putting Together the Wheel with Yin and Yang Energy
37

PART IV
Astrological Houses and Sign Keynotes
47

PART V
Astrological Calculations: The Importance of Accuracy
Understanding the Mechanics of Astrology and the Charts
75

About the Author
Eleanor Haspel-Portner, Ph.D.
107

Acknowledgments
109

Illustrations

Illustration 1: The Noble Sciences Mandala 9

Illustration 2: The Body Energy Map with Centers Labeled 12

Illustration 3: A Basic Noble Sciences Body Map 14

Illustration 4: Yang/Yin 1 becomes 2 becomes 4 16

Illustration 5: The Mandala of Synthesis 17

Illustration 6: Astrology Signs: Color Coded with their Names 18

Illustration 7: The Astrology Wheel Used in the Mandala of Synthesis:
        An Integration of Science and Intuitive Knowledge 27

Illustration 8: Significant Years Activated by Planetary Cycle 29

Illustration 9: I-Ching Yang/Yin Image in the Noble Sciences' Wheel 33

Illustration 10: Table of Fast-Moving Planets
        Planet, Symbol, Keyword, House, Sign 36

Illustration 11: Table of Slow-Moving Planets
        Planet, Symbol, Keyword, House, Sign 36

Illustration 12: Mandala Showing Yin/Yang in Wheel 39

Illustration 13: Yin 39

Illustration 14: Yang 41

Illustration 15: The Astrology Wheel in the Mandala of Synthesis:
        An Integration of Science and Intuitive Knowledge 41

Illustration 16: Table of Astrological Sign, Planet, House 42

Illustration 17: The Colored Astrological Wheel with Signs, Planets, and Houses 43

Illustration 18: Astrological Wheel with Quadrants Labeled 45

Illustration 19: The First Quadrant Houses 49

Illustration 20: The Second Quadrant Houses..50

Illustration 21: The Third Quadrant Houses 51

Illustration 22: The Fourth Quadrant Houses...52

Illustration 23: Astrological House Keynotes and Areas of Life Affected ...53

Illustration 24: Planets in the Wheel in the Houses They Rule 55

Illustration 25: Astrology Planet Keynotes: Experience & Behavior 57

Illustration 26: Trines of Elements 59

Illustration 27: The Fire Elements 61

Illustration 28: The Earth Elements 62

Illustration 29: The Air Elements 63

Illustration 30: The Water Elements..64

Illustration 31: Mandala Cusps and Midpoint Trines 65

Illustration 32: Element Triangles 66

Illustration 33: Qualities: Squares..67

Illustration 34: The Cardinal Squares and Oppositions...............................69

Illustration 35: The Fixed Squares and Oppositions 70

Illustration 36: The Mutable Squares and Oppositions................................71

Illustration 37: Qualities: Squares, Cardinal, Fixed, Mutable 73

ILLUSTRATIONS ♦ 3

Illustration 38: Mandala Cusps and Midpoint Squares 74

Illustration 39: The Astrological Clock 77

Illustration 40: East-North-West-South in the Astrological Wheel 78

Illustration 41: Telling Time in the Astrological Clock 79

Illustration 42: Time and Space in the Astrological Wheel 81

Illustration 43: Directions, Time, and Orientation 83

Illustration 44: Reviewing the Wheel 86

Illustration 45: Latitude and Longitude 88

Illustration 46: AJS February 19, 1992, 6:53 am, San Antonio, Texas –
How time affects an astrological chart 90

Illustration 47: AJS February 19, 1992, 6:56 am, San Antonio, Texas –
How time affects an astrological chart 90

Illustration 48: AJS February 19, 1992. 6:53 AM, San Antonio, Texas –
How time affects a Body Graph 91

Illustration 49: AJS February 19, 1992, 6:56 AM, San Antonio, Texas –
How time affects a Body Graph 91

Illustration 50: AJS February 19, 1992, 6:56 AM CST, San Antonio,
Texas – How place affects an Astrological Chart 92

Illustration 51: AJS February 19, 1992, 6:56 AM EST, Charlotte, North
Carolina – How place affects an Astrological Chart 92

Illustration 52: AJS Body Energy Map – February 19, 1992, 6:56 AM
CST, San Antonio, Texas – How place affects the
Body Graph 93

Illustration 53: AJS Body Energy Map – February 19, 1992, 6:56 AM
EST, Charlotte, North Carolina – How place affects
the Body Graph 93

Illustration 54: Latitude and Longitude 95

Illustration 55: Blossburg, Pennsylvania, July 22, 2008, 12:00 PM EDT,
Latitude: 41°N40'44", Longitude: 077°W03' – How
Latitude and Longitude affect an astrological chart 96

Illustration 56: Lima, Peru, July 22, 2008, 12:00 PM EDT,
Latitude: 12°S03' Longitude: 077°W03' – How
Latitude and Longitude affect an astrological chart 96

Illustration 57: The Astrological Signs converted to degrees
in the Astrological Clock 99

Illustration 58: An Astrological Chart Example 101

Illustration 59: Color Key 102

Illustration 60: Astrology Legend 103

Illustration 61: Table of Planetary Movement 104

Illustration 62: Hexagram Correspondences to Zero Degree Points
of Each Zodiac Sign 105

Legend: 105

Foreword

If life is about learning lessons and achieving mastery over various stages of development, then the goal of Noble Sciences is to provide information and guidance for you in your own achievement of these stages, preparing you to succeed as you enter each new cycle.

All books in Noble Sciences Basic Tools Series contain foundational material for use as part of the complete Noble Sciences Sacred Synthesis. This synthesis has its roots in components of The Kabalistic Tree of Life, Hindu Chakras, Western Astrology, The I-Ching, the Human Design System, and Interdisciplinary Human Development. Noble Sciences Synthesis breaks new ground in its comprehensive integration of social scientific and esoteric knowledge.

Noble Sciences' work combines and integrates all of these systems in very specific ways. The complexity of Noble Sciences lies in its recognition and documentation of the multidimensional structural dynamic of consciousness. You need years of study in the roots of Noble Sciences' disciplines to master them. However, to properly utilize Noble Sciences' tools so you can use them productively, you are well served by having a basic understanding of each of them. In this way, you have enough information to recognize the breadth and scope included in the unique work of Noble Sciences.

As the creator of Noble Sciences, a clinical psychologist, a social science professional, and a student of esotericism, I am deeply committed to the marriage of both the scientific and the esoteric. It is my goal to combine them without losing their essence or the integrity of the individual in the process. Thus, no material in Noble Sciences is ever formulaic. To properly use the knowledge that is at its base, it is essential to learn the underlying concepts and stretch your mind to grasp the complexity involved so you can begin to decode the multidimensional world for yourself, in your own way.

In Loving Light,

Eleanor Haspel-Portner, Ph.D.
Pacific Palisades, California

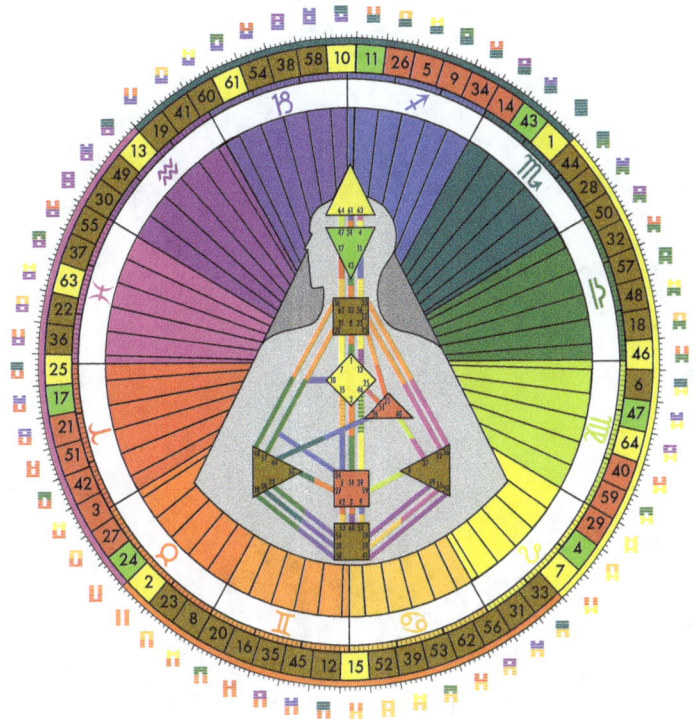

PART I

NOBLE SCIENCES
SACRED SYNTHESIS

Components

Noble Sciences' work synthesizes information based on the Kabalistic Tree of Life, the Hindu Chakra system, Astrology, the Chinese I-Ching, Human Design, and Developmental Psychology. The Noble Sciences Mandala shows the way the information is coded into a body map and is ready for de-coding.

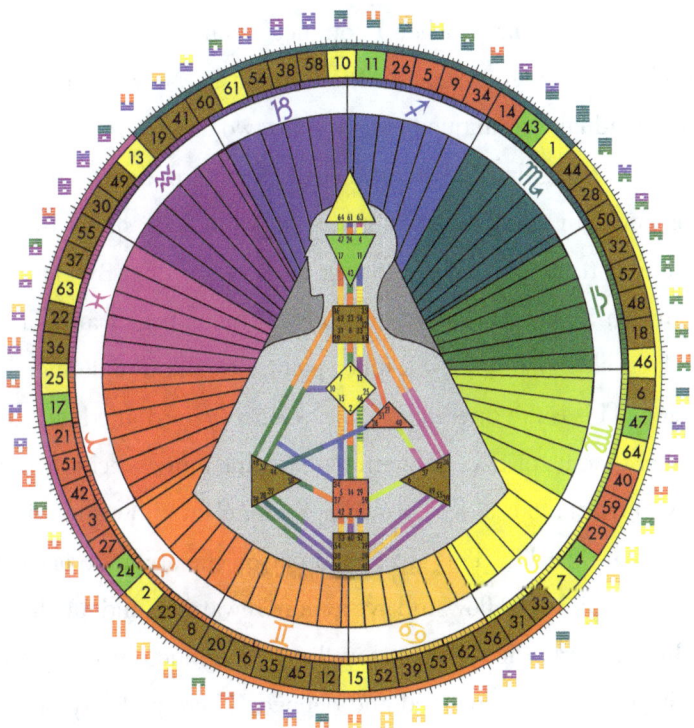

Illustration 1: The Noble Sciences Mandala

The Human Design System, created by Ra Uru Hu, synthesized the components of Kabalah, the Chakras, Astrology, and I-Ching into a body graph and recognized the way the information was coded into the human at the time of birth and prior to it. By noting the prenatal influence of the planets and their role in how a person functions energetically, Human Design opened the door to more research and documentation that confirmed the importance of looking at an individual in multiple time frames both in terms of the energy of their own consciousness and in determining how they actually function in their manifesting life.

As a professional psychologist and social scientist who also studied Astrology, the Tree of Life, and the I-Ching, I was both fascinated by the charts and dismayed when I conducted statistical research on the basic Human Design system and found that it was very incomplete. Each individual is a unique being and no two charts are exact. As an astrologer, I recognize the importance of exact birth data, as described in this book. In addition, developmental psychology shows that critical periods track how a baby develops through time. By expanding the Human Design research to include postnatal time frames as well as additional prenatal ones, a fuller picture of the unique individual emerged. This is the data that Noble Sciences contributed and coded into the Mandala of Sacred Synthesis, creating more precision and accuracy in chart analysis and providing insights into how an individual functions in their life. This gives each person the ability to learn how to be in the Universal flow and live a more rewarding and fulfilling life.

In many esoteric systems, as well as in acupuncture, energy pathways in the body are mapped and defined. These pathways are generally thought of as flowing patterns of energy in the body anchored by hubs that are called chakras, Sephiroths, centers, flows, and channels, among other things. In my introductory book, "Beyond Human Design: Cosmic Secrets Revealed" I introduce these concepts in the Noble Sciences work. You can also read more about energy and its centers by exploring my work at www.noblesciences.com or www.beyondhumandesign.com.

In working with the mandala, it is important to note that there are correspondences in the Noble Sciences system to the Tree of Life, but there are also variances. Notice that the central image in the mandala shows nine energy centers within the body map. These mostly correspond with the centers (or "Sephiroths") designated in the Tree of Life, which actually hold ten energy centers in its system. These energy centers all contain specific meanings and functions, and have 32 pathways (or Paths of Intelligence), which represent unfolding consciousness in the various worlds. The basic structure and progression of development in the process of individuation, or evolving consciousness described by psychologists, follows the same pathways as described in the Tree of Life. This progression unfolds in a specific sequence in order for development to advance and for each individual to reach realization and self-actualization.

In the tradition of the Tree of Life, there are many layers to the evolutionary current of light, which brings all physical manifestation into existence. These layers reside in the void and when light first forms in the space of the void, the number one comes into existence. Created from this are the twelve signs of the zodiac, seven planets, and four elements. Thus, the Tree of Life folds into itself like a hologram, containing all of the information held in each particle of the universe.

The Hindu Chakra system uses seven, rather than nine energy centers. However, Noble Sciences' body map contains these centers, plus two more, and, in fact, differentiates them more completely. Please note that there is not an exact correspondence between the energy centers in Noble Sciences, the Tree Of Life, or the Hindu chakra centers. The additional centers provide more detailed information about how individuals process their world. The Splenic Center illustrates how we can intuitively understand the health of our physical body. The Solar Plexus deals with our emotions, desires and beliefs, and the Heart Center is a physiological hub that represents the actual heart itself and controls the breath.

Look at the following Body Energy Map that Noble Sciences uses.

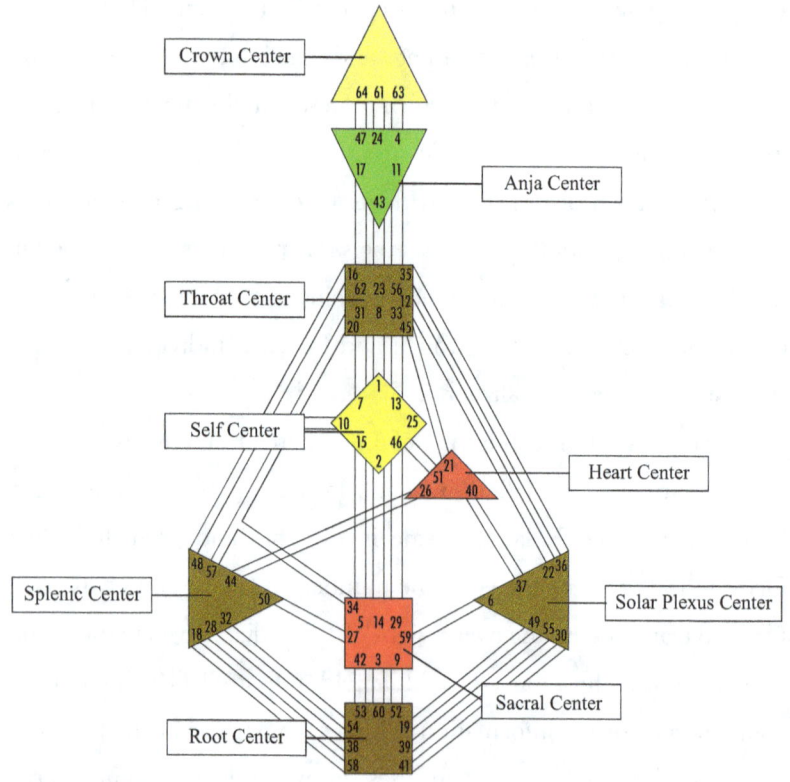

Illustration 2: The Body Energy Map with Centers Labeled

In Noble Sciences, the energy centers work like this:

- **THE ROOT CENTER** grounds us, providing our fundamental support.

- **THE SACRAL CENTER** contains our instinctive gut feelings, life force energy, and is the source of kundalini.
- **THE SPLENIC CENTER** is body oriented. It deals with our physical and biological functions, informing us about the state of our health on a subconscious physical layer, and is responsible for the operation of our immune system.
- **THE SOLAR PLEXUS** is our emotional center. It's related to feelings, translating them into awareness for conscious interaction.
- **THE HEART CENTER** deals primarily with the biological layer and the actual physical heart itself, and thus, functions at a deep unconscious level of our physiology. It also carries the energy of will power giving us dominion over our energy and our capacity to monitor our breathing and stress.
- **THE SELF CENTER** contains the core of our identity. It holds us together in time and space. It serves as a link between the lower energy centers and the higher ones (or the "supernals"). This is where we're able to experience our Divine Self, Higher Self, and Integrating Self.
- **THE THROAT CENTER** drives our ability to bring things into manifestation and is related to self-expression.
- **THE AJNA CENTER** is a switching station. It is the center for creative intelligence and central cognition. It grounds ideas in reality by helping us create meaning, and helps us translate gut feelings into higher truth and knowing.
- **THE CROWN CENTER** is a pressure center that gives us the desire to learn, understand, and make sense of things.

Gates and Pathways

Gates and Pathways
Prenatal Date   Natal Date
34.3 ☉ 55.1
20.3 ⊕ 59.1
39.4 ☽ 64.6
38.1 ☊ 58.6
39.1 ☋ 52.6
11.1 ☿ 37.1
57.3 ♀ 60.5
14.3 ♂ 60.5
64.2 ♃ 40.6
41.1 ♄ 19.5
38.3 ♅ 54.2
38.6 ♆ 54.4
43.3 ♇ 43.5
33.3 ⚷ 31.4

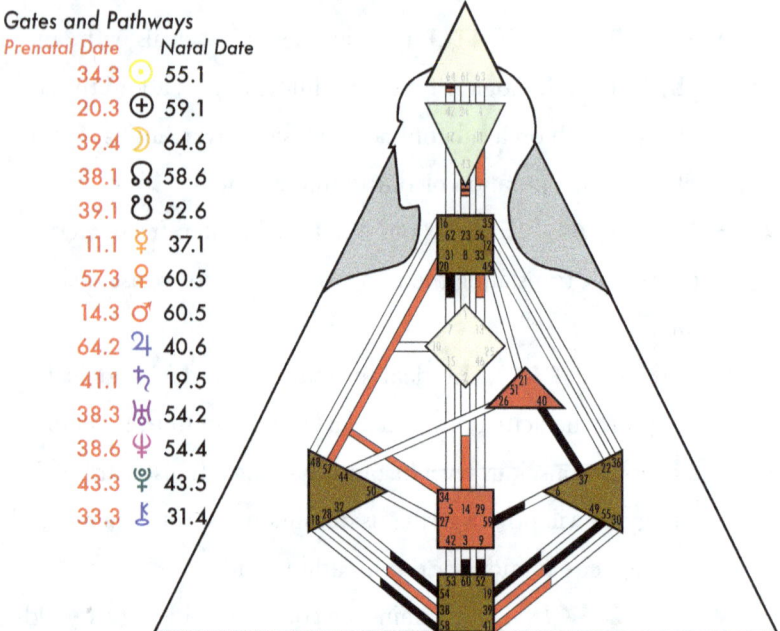

Illustration 3: A Basic Noble Sciences Body Map

There are 64 active energy gates in humans. Each gate has a meaning because it corresponds to a hexagram, and each of these contains the values of masculine or feminine, as well as the elements and worlds they're related to. The worlds are: the mental, spiritual, emotional, and physical worlds. These worlds encompass all aspects of how we understand, learn, and grow from experiences in our life. They are represented as physical forms, archetypes, symbols, and interpretations of reality.

The channels on the body map above are interpreted as follows:

- Those channels that aren't colored in, or remain white, are undefined or neutral. They receive all activating signals without predisposing filters.

- Those channels that are black contain active or activating energy. They slant responses and perceptions in certain ways.
- Red channels contain passive or transforming energy. They operate below the level of conscious awareness but are predisposed to certain ways of responding.

When a gate is defined, it is active, and gives more clarity to decisions that need to be made. However, when a gate is inactive, all options are available, and the individual has many things to choose from. For example, if the Self Center isn't defined in your chart, you would take all information into your system, without clarity as to which direction to go. Individuals with undefined or inactive Self Centers are vulnerable to the needs of others and tend to be more selfless, confusing their own needs with those of others.

As in the Tree of Life, the Body Graph, or Map, describes multiple layers and interpretations determined by the world under consideration. In the Archetypal world, the world of the collective unconscious that connects humans with all other living beings, there are hypothesized to be three realms that activate in any human. Archetypes are described in terms of realms: the realm of survival; the realm of earth plane reality; and the realm of the light field of transcendence. These are activated during sleep and dreaming in the human.

In many respects, Noble Sciences' astrological charts look like standard diagrams, with signs located on the outside of the wheel. The houses within this 360-degree wheel are divided into 30-degree intervals. In this book, the astrological components of the wheel are laid out so you can understand the way to read the signs and planets as they affect and influence an individual by their movement. Since a human being is a living entity, it is essential to recognize that no static system can describe an individual. Instead, it is important to take into account and understand that the planetary movements influence an individual in complex ways and in accord with where and how they impact an individual's unique chemistry.

Another component of Nobles Sciences' charts and interpretation is the Chinese I-Ching, which is an ancient tool that was used for divination. In recent years much research into the I-Ching has been done. One hypothesis is that the I-Ching contains a binary code of the genetics of humans. There are 6 lines to each Hexagram in the I-Ching and there are six key amino acids that combine in forming the genetic code of an individual.

In addition, each Hexagram consists of two trigrams, each trigram having three lines that are either yin or yang. A yang or solid line is combined with another line that may be solid or broken. A broken line is yin. From these two lines, four combinations are possible. And when a third line is added, each yang line can have four combinations, and each yin line can have four combinations as shown in the illustration.

Illustration 4: Yang/Yin 1 becomes 2 becomes 4

If you put the 32 paths (which are similar to channels) that connect with the various energy centers in each of the twelve 30 degree astrological signs, you get the 384. This is the same number as the number of lines in the 64 corresponding I-Ching hexagrams. In another book, I go into detail about the numerology and layout of the I-Ching in the Mandala, but for the purposes of this document, note that each hexagram corresponds to a range of degrees in the astrological wheel and the numbers along the outer rim of the wheel denote the Hexagram to which it refers.

Illustration 5: The Mandala of Synthesis

No two charts are the same. To learn to read a chart you must be prepared to study each chart for its unique energy configuration and use the information coded into it in its interactive nuances, in the same way an individual experiences these components.

Enjoy your study. It is deep and exciting and will yield great insights into yourself and others. However, be aware that the truth rests with the individual and not the interpretation. Choice is always possible and never predetermined in a chart. What an individual decides to do with their energy is left for them to decide.

The following Illustration shows a basic astrology chart with the names of the Astrological Signs and their Symbol, also known as their Glyph. In learning the language of astrology, it is important to learn the names of the signs and their symbols.

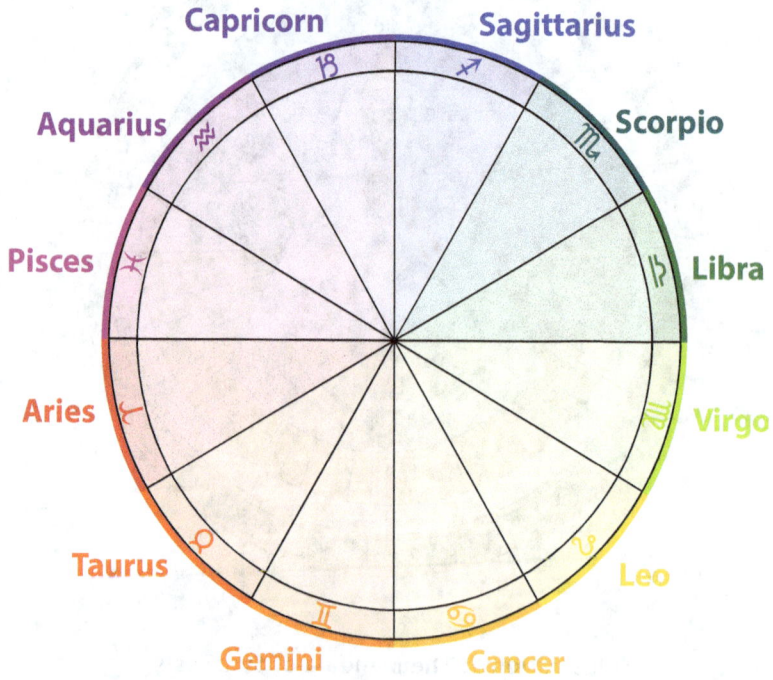

Illustration 6: Astrology Signs: Color Coded with their Names

Note: All Noble Sciences materials are color-coded to correspond to the colors used in the Golden Dawn System, teachings from an esoteric study that involved Astrology, Alchemy, and Theosophy.

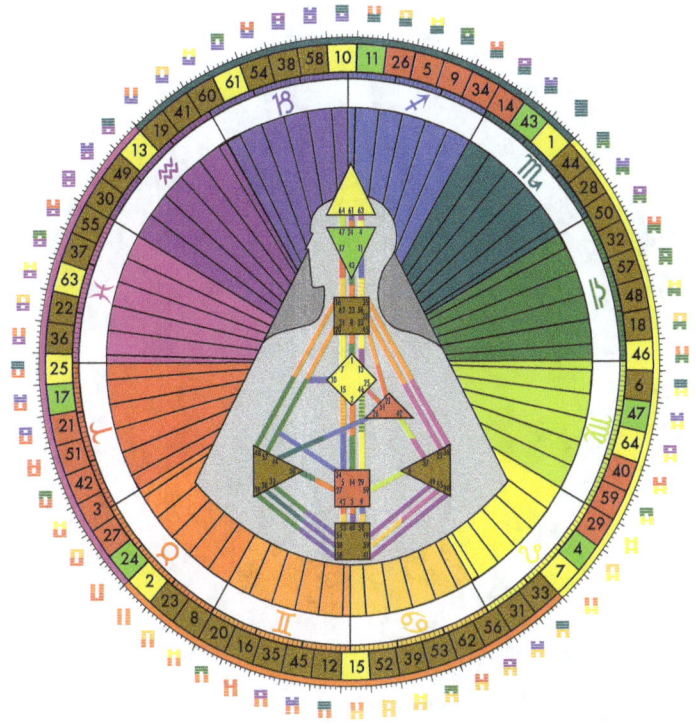

PART II

ASTROLOGY FOR NOBLE SCIENCES

What You Need To Know

Noble Sciences uses astrological information and other systems in calculating a map of how someone functions. A basic knowledge of astrology and some underlying concepts involved in calculating and reading a chart is important for a full interpretative reading of a Noble Sciences Chart.

Like many other disciplines with roots in ancient times, Astrology is a complex study that requires a basic understanding of the way charts are calculated. This includes understanding why the calculations require exact birth times, which goes a long way in putting the knowledge of the charts into perspective.

Noble Sciences charts go Beyond Human Design knowledge. Thus, it is important to read the coded information in its planetary and structural nuance based on both the Tree of Life and on the astrological data contained therein. This book gives the reader all the important essentials for reading planetary nuances without the necessity of studying astrology in depth.

In *Astrology Essentials* you will learn:

- The scope of Astrology: its historical perspective, scientific basis, and its interpretive art.
- Why time and place are important in calculations for Noble Sciences.
- How the Astrology Wheel and the Mandala are organized: why cycles are important in Human Development.
- The Astrology Alphabet – its Symbols including:
  - Quadrants and the Wheel

- Houses – what they are, how they're numbered, keywords associated with planets and signs
- Planets – what they are, keywords, their timing, associated house and sign
- Signs – what they are, keywords, and their elements, associated houses and planets
- The Way the astrological alphabet works together
  - Elements (fire, earth, air, and water)
  - Qualities (cardinal, fixed, and mutable)
  - Houses, planets, and signs, in combination with the elements and qualities
- How to read interactions – the art of interpretation, i.e., putting the alphabet together.

The Art of Interpretation in Noble Sciences

Astrological charts use exact birth data, which includes date, time, and exact location. The accuracy of the data is essential in calculating correct charts/maps.

Planetary positions are noted for their sign, house position, element, quality, and rulership (how comfortable they are in their house position with its native characteristics, sign, degree position, element, and relationship to other planetary positions). These characteristics are all considered as they integrate in the Noble Sciences synthesis including their positions in the body energy system.

Multiple calculations used in Noble Sciences show layers of consciousness that makes each individual's functioning unique.

The History of Astrology

The roots of Western Astrology date back to Babylonia, evolving over thousands of years to the present day. In ancient times, when people looked at the sky, they saw certain patterns there. After time, they realized that these patterns corresponded with events that happened on Earth, as well as in the lives of individuals, and began to look at these patterns as celestial signs.

They noticed that certain patterns appeared to have increased probabilities. Repeated observation indicated that certain "signs" or times of year associated more with specific planets and types of events than others.

In ancient times, it was noted that particular celestial or planetary bodies pointed to different astrological positions, or "signs." The dates related to these "signs" became the basis of astrological symbolism and of what we call "zodiac signs."

Astrological wheel divisions (houses), signs, and their planetary geometry pointed the way toward evaluating and understanding numerous things. Ongoing research expanded observations of planetary bodies to include an increasing diversity of events and life areas.

As astrological information became increasingly accepted, astrologers studied internal relationships, personal relationships, and interpersonal relationships. They observed relationships between planetary bodies as likely to influence people in particular ways. Over time many identifiable patterns proved to correlate with life events.

In recent times, astrology has become increasingly accepted as more scientific and less belief-based, and is gaining respect as more than pure fortunetelling. Sun sign astrology is the kind of astrology published in newspapers and magazines for the general public and has become recognized as superficial, less precise, and less scientific than more extensive accurately calculated charts. As a professional psychologist who also taught college classes in psychological testing, I am qualified to say that I have found that in the hands of a good astrologer, a detailed astrological chart is as powerful a tool for analyses as psychological testing.

Astrological Analysis at Its Best

Computer Astrological Mathematical Calculations produce volumes of data. These precise tools for astrological prediction and analysis overwhelm many people. Nevertheless, analyses from such data can be reproduced and validated. This approach is scientific.

Intuitive astrology abandons mathematical precision. It often uses astrological information without a depth of technical understanding. Those using purely intuitive tools "read what feels right." Scientifically oriented interpretations are based on replicable patterns and thus, are likely to be made by scientifically trained astrologers.

Neither the scientific nor the intuitive approach is in error. Scientific and intuitive astrology tools are both valid and valuable. When both approaches join together in analysis, the depth of evaluation reaches beyond the limitations of either tool alone.

Science and Intuition in Astrology

Perspective and Orientation of the Zodiac shows the planets in relationship to the Sun and to the Earth. They are important for accurate astrology chart calculations.

Historical roots broaden understanding because astrology grew out of both science and intuition. It began with the scientific process of hypothesis testing and observation with its roots in mathematics (scientific) and esoteric/metaphysical astronomy (intuition). Numbers and logic became the foundation for astrological calculations.

From this combination of science and intuition patterns of astrological configurations emerged that became the art of astrological interpretation. To have a true synthesis that tapped the human potential of chart analysis some very specific guidelines have to be followed in calculating charts for them to be accurate and to provide the astrologer with reliable data.

Your view or perspective of the Sun and planets depends on your exact location on Earth. When you watch a beautiful sunrise in London, your view of the sky is different from what your view would be if you watched an equally beautiful sunrise in Hawaii. Although the planets are in the same formation, they appear differently from different locations on Earth at that time. Your view or perspective of the Sun and your perspective of the planets change with your location. This explains why an eclipse is visible in one location but not another.

Thus, time and location are both important. Accuracy depends on exact data. Your charts are only as accurate as the information provided for their calculation.

Much in the way that a scientist has an unconscious effect upon that which he's observing, planets affect our lives and events once we discover them and then become part of our conscious inclusion. For example, Uranus did not have an influence until its discovery in 1781 and then it became something of a cosmic alarm clock, sparking the

Industrial Revolution, and participating in the American Revolution in relation to the concept of freedom.

Cosmic shifts in consciousness also seem to have occurred with the discovery and awareness of Neptune (September 23, 1846 in Berlin) and of Pluto (February 18, 1930 in Flagstaff, AZ)..

Neptune was first observed in conjunction with Jupiter in 1612, but at that time, Galileo misidentified it, thinking it was one of the Jovian moons. From 1834 to 1846, T. J. Hussey, George Airy, John Adams, and Urban Le Verrier all observed Neptune, but they failed to identify it as a planet. It was finally Johan Galle, who with the guidance of Le Verrier, correctly identified Neptune on September 23, 1846 at 10 pm GMT. (The birth of the Outer Planets by David McCann in Skyscript.co.uk)

Just as the discovery of Uranus was filled with surprise, the discovery of Neptune was filled with confusion. Neptune and Pluto were both in retrograde when they were discovered, and this fact reflects the way they generally manifest in a person's life, through unconscious experience rather than in conscious reality experiences. At the time Neptune was observed and discovered, many confusing and difficult world events were occurring, including the rise of Communism and growth of Mormonism.

That said, there is nothing good or bad in any astrological pattern. The expression of the planets and how you use them is your choice. The consciousness you bring to every situation is there for transcendence.

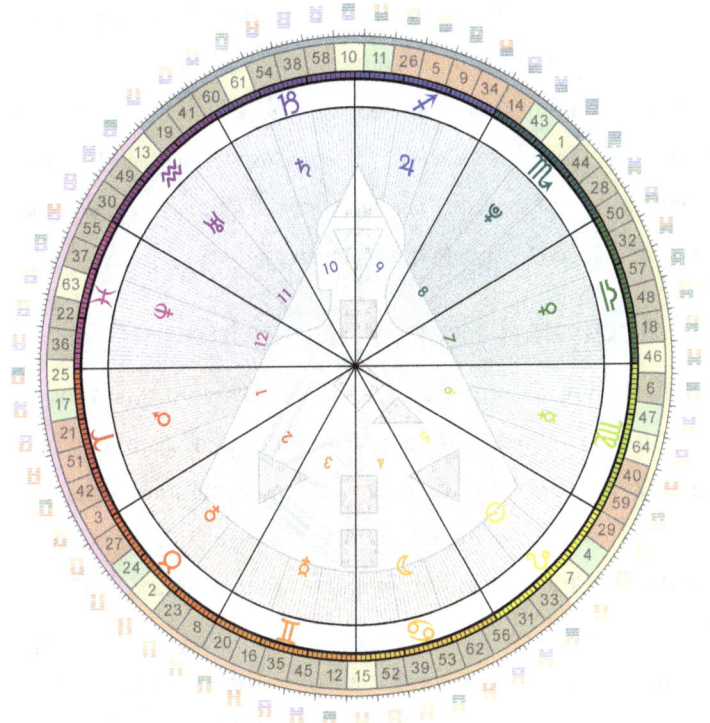

Illustration 7: The Astrology Wheel Used in the Mandala of Synthesis: An Integration of Science and Intuitive Knowledge

Since a circle is 360 degrees, it is divided into 12 thirty degree sections that form the astrological zodiac with each 30° section representing one zodiac sign. (360°).

TIME moves around the wheel in a **clockwise direction**. The wheel is divided into 24 hours (equal to one day), with each interval representing two hours.

HOUSES move in the opposite direction from time, in a **counterclockwise direction**. The first house to be counted is Aries (9 o'clock on the time clock we use), which is also where your ascendant is located (your rising sign). Where the planets land within the wheel, houses, and degrees, determine the context, significance, and meaning they express in an individual's chart.

Noble Sciences color-codes all components of its charts in order to enhance connections and clarify interpretations. Understanding the components in the wheel, what they mean, and how they integrate, involves learning something about the dynamics of the charts' components and how they operate.

Astrology and Human Development

Astrology is a complex discipline. Its wheel shows transpersonal human development throughout the life span. Astrology analyzes the zodiac and combines a system of house divisions, signs, planets, their complexities, and nuances of their interrelationships. As planets move through the zodiac they impact different areas of life globally and personally.

The houses and planets in their signs move around the wheel in sequence. The sequence of development visible in astrological charts influences events in numerous ways. Planets follow regular patterns that influence specific life passages at certain times.

An Individual Moves Through Different Stages At Predictable Times

Take some time looking at the chart of planetary cycles and consider how planetary activations impacted you at significant years in your life.

Age	(♃) Jupiter	(♄) Saturn	(♅) Uranus	(♆) Neptune	Age	(♃) Jupiter	(♄) Saturn	(♅) Uranus	(♆) Neptune
3	✓				45	✓			
6	✓				48	✓			
7		✓			49		✓		
9	✓				51	✓			
12	✓				54	✓			
14		✓			56		✓		
15	✓				57	✓			
18	✓				60	✓			
21	✓	✓	✓		63	✓	✓	✓	
24	✓				66	✓			
27	✓				69	✓			
28		✓			70	✓	✓		
30	✓				72	✓			
33	✓				75	✓			
35		✓			77		✓		
36	✓				78	✓			
39	✓				81	✓			
42	✓	✓	✓	✓	84	✓	✓	✓	✓

Illustration 8: Significant Years Activated by Planetary Cycle

Many developmental stages depend on one another, e.g., a baby only walks after learning to crawl. The next level of development is attained only after the first is mastered. There is a sequence that must be followed and each moment defines the next, like a spiral.

Once a stage develops, it reaches balance (equilibrium) and a new cycle begins. It creates imbalance (disequilibrium) as it evolves toward a more differentiated level and the cycle begins again. Thus, life-stages generally follow a pattern that increases mastery of life and its challenges.

The Date, Time, and Place Calculations in Noble Sciences Maps

Each Noble Sciences Chart/Map uses two timeframes in its calculations—the moment under consideration, and either the prenatal or postnatal calculation. A standard astrological chart uses only one timeframe for analysis. Noble Sciences uses a six-month timeframe—three months before birth to three months after—to address how the energy of the planets influence the current incarnation and reveal the intelligence within creation.

- From conception through the first three months of life outside the mother's body, one full year of human development takes place, with the last three months after the date of birth composing the missing fourth trimester. Within that year, the universal macrocosm is internalized so that by the end of the first three months of life, the infant has gone through a whole cycle of planetary rotations of the sun and earth, and is influenced by all of these things in their consciousness.
- By the end of the first trimester, the baby has developed all nine energy centers. The sacral and head centers are the last to be developed during this time.
- At the end of the second trimester, the baby's physical features are more complete and its energy centers are beginning to function. Now consciousness is ready to activate. The upper and lower worlds within the infant are linked through the personality crystals located at the Ajna center inside the head, and in the Sacral center located in the pelvis.
- During the third trimester or prenatal stage, consciousness is activated.

- At one week before delivery, there's a shift in consciousness. New connections form as hormones mobilize.
- The emotional body is energized two days before birth. This moment is sometimes connected to the pre-labor contractions (known as Braxton-Hicks contractions) that some women experience.
- At one hour before delivery, the physical layer of the infant is stimulated.
- Birth marks the beginning of the missing fourth trimester. The infant starts to respond to the outside world at this time.
- One week after birth, the spirit is anchored in the body. The baby's spiritual body integrates with the emotional and physical bodies. It is a strong unifying force in development.
- By the end of the fourth trimester, or postnatal stage, the baby begins to exert will, acting with volition, reaching and interacting with the physical world and those around him.

Development throughout life moves in a spiraling pattern through all layers of awareness: mental, spiritual, emotional, and physical, and each layer of awareness follows its own timing, pattern, and movement. There are cycles within these stages of development between the third trimester and three months after a baby is born. They are as follows:

MENTAL: This developmental calculation is based on the Sun rotation. It marks three months before birth and continues until three months after birth when awareness, imprinted with outside planetary influences, activates volition, or the ability to make conscious choices.

SPIRITUAL (OR ARCHETYPAL/UNCONSCIOUS): This developmental calculation is based on the Moon rotation. It begins one week before and continues until one week after birth. It shows the timing of the unifying integrating aspect of consciousness that connect our unconscious to our conscious mind during Rapid Eye Movement (REM) dreaming.

PHYSICAL: This is related to the sleep matrix and begins an hour and a half to two hours before or after birth. The physical body at this level functions without conscious intervention.

EMOTIONAL: This is initiated four days before birth and continues until four days after birth. It relates to human connectedness, parent/child attachment, and bonding as well as other emotional reactions.

INTEGRATIVE: When all worlds or chart calculations are combined they show the way a person organizes their energetic bodies as a whole and actually function in the world. It is essential to include all layers of awareness into this picture of a human being.

Human consciousness is complex. The Mental world is the reality world in which we process information and it is informed by our daily life as well as by the unconscious components of the other worlds. The Unifying Integrative Field of Consciousness (**REM**) connects the deep collective unconscious to the superconscious higher Self. It is activated about an hour and a half before birth connecting the physical to the spiritual body. The unconscious worlds process information for the benefit of the emotional and mental bodies, almost instantly and often without conscious awareness. Sleep and dreaming are included in the integrative awareness because they are essential for health and restoration of the body and psyche. It is this integrative function that allows living beings to function as a unified whole.

The developmental process continues throughout life and is reflected in the ancient concept of Yin and Yang. When harmony is achieved by the two forces of energy, a level of development has been mastered, and the energy is re-set for the next stage of development to begin.

The I-Ching Yang/Yin in Noble Sciences Charts

Noble Sciences integrates the I-Ching method of divination as represented by the Yin and Yang energies within its astrological charts and maps. This provides a deeper analysis and understanding for the development, emotional processing, and events in an individual's life.

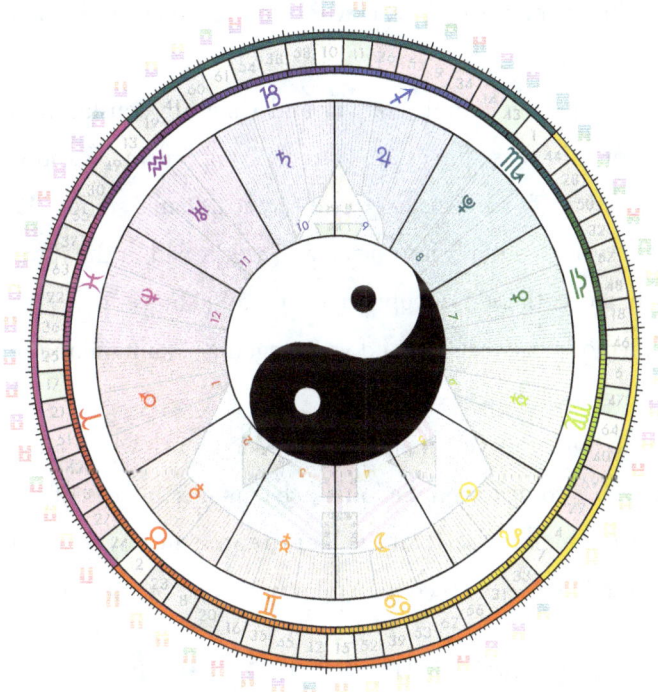

Illustration 9: I-Ching Yang/Yin Image in the Noble Sciences' Wheel

- Yang is White. It represents the masculine, the sun, action, initiating, daylight, warmth, and strength. Yang energy is strongest at 45° of Scorpio ♏, and moves around the wheel in balance to yin or receptive energy. Yang energy is weakest at 45° of Taurus ♉.
- Yin is Black. It represents the feminine, the moon, passivity, receptivity, coolness, night, and intuition. Yin energy is strongest at 45° of Taurus ♉, and moves around the wheel in balance to yang or activating energy. Yin energy is weakest at 45° of Scorpio ♏.
- Energy moves both clockwise and counterclockwise along the wheel.

The opposing energies of yin and yang are constantly shifting. In spite of this, these forces are actually complimentary and vital for bringing everything in existence into physical manifestation. When in balance, everything falls into place and works within the flow. When out of balance, there is disruption, chaos, and illness.

Harmony between yin and yang together maintain and produce everything in existence.

As in human development, a new cycle is born when a cycle ends or a stage has come to completion. Each contains the seed of the other. Within the Yin, is the seed for the Yang (represented by the white dot). Within the Yang, is the seed for Yin (represented by the black dot).

Astrology Cycles in Human Development

Planetary cycles facilitate transformation and mastery throughout life. Spiraling life development touches mental, spiritual, emotional, and physical layers of life while astrological time cycles follow developmental patterns around the wheel. They balance active internal development, receptive internal development, receptive worldly development, and active worldly development. Planetary movements follow zodiac wheel divisions of houses and their progressions.

Once balance is achieved in one area, it begins a new cycle. It creates disequilibrium as it evolves and moves toward a more differentiated level of equilibrium. Sometimes this occurs quickly and sometimes it can take a long time, depending on whether slow-moving or rapid-moving planets activate sensitive planetary configurations in the chart.

The fast-moving planets are the Sun (☉), Moon (☽), Mercury (☿), Venus (♀), Mars (♂), and sometimes Jupiter (♃). These planets point to events and changes that take place in an individual's life as part of their personal process. When they trigger an astrological pattern in an individual's chart they result in the commencement and type of event in a person's life.

The Sun and Earth always seek balance because they're always 180° apart, that is, they are in astrological opposition(☍).

The slow moving planets are, Saturn(♄), Uranus (♅), Neptune (♆), Pluto (♇), Chiron (⚷), and the nodes (☊☋) (although technically, the nodes aren't planets—they're a mathematical equation that has a tangible impact within a chart). These planets can take several generations to move through a cycle. They indicate transpersonal major life-turning points and represent global or collective processes for whole generations.

Planet	Symbol	Keywords	House	Sign
Sun	☉	Vitality, Individuality, Spirit, Will	5th House Fire/Fixed	Leo ♌
Earth	⊕	Grounding, Foundation, Roots	2nd House Earth/Fixed	Taurus ♉
Moon	☽	Emotional Feelings, Attachments, Children, Instincts, The Feminine, Receptivity, Nurturing	4th House Water/Cardinal	Cancer ♋
Mercury	☿ ☿	Communication, The Mind, Links to Spirit and Matter, Short Trips	3rd House Air/Mutable	Gemini ♊ Virgo ♍
Venus	♀ ♀	Harmony, Beauty, Affection, Relating, Aesthetics	2nd House Earth/Fixed	Taurus ♉ Libra ♎
Mars	♂	Initiative, Energy, Courage, Action, Assertiveness	1st House Fire/Cardinal	ARIES ♈

Illustration 10: Table of Fast-Moving Planets
Planet, Symbol, Keyword, House, Sign

Planet	Symbol	Keywords	House	Sign
Jupiter	♃	Expansion, Good Fortune Positive Growth	9th House Mutable/Fire	Sagittarius ♐
Saturn	♄	Limitation, Discipline, Effort, Constraint	10th House Cardinal/Earth	Capricorn ♑
Uranus	♅	Originality, Independence, Change, Invention, Unexpected	11th House Fixed/Air	Aquarius ♒
Neptune	♆	Compassion, Sensitivity, Confusion, Imagination, Loss, Chaos	12th House Mutable/Water	Pisces ♓
Pluto	♇	Deep Transformation, Regeneration, Coercion	8th House Fixed/Water	Scorpio ♏
Chiron	⚷	Healing, Consciousness, Awareness, Expansion	9th House Mutable/Fire	Sagittarius ♐
North/South Node	☊☋	Relates to Destiny and Life Purpose, Past and Future		

Illustration 11: Table of Slow-Moving Planets
Planet, Symbol, Keyword, House, Sign

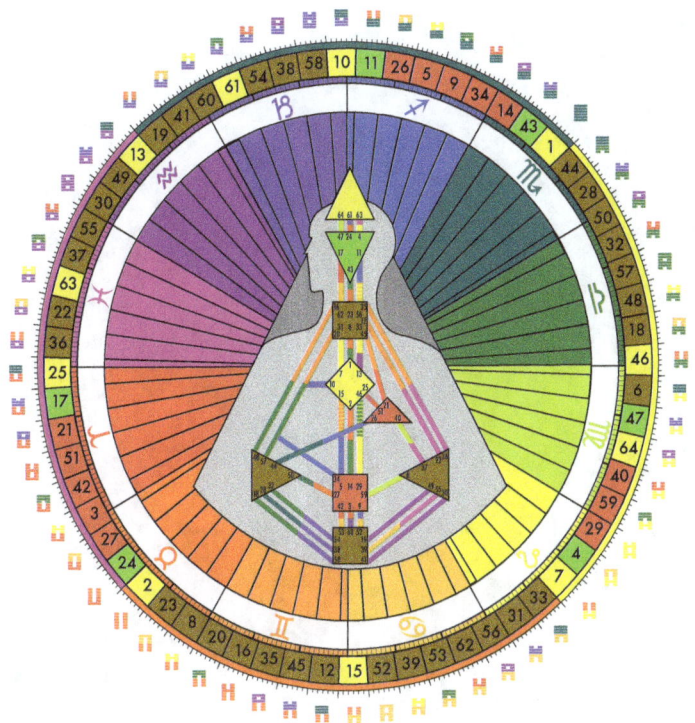

PART III

PUTTING TOGETHER THE WHEEL WITH YIN AND YANG ENERGY

Planets impact the life cycle through timing and house position

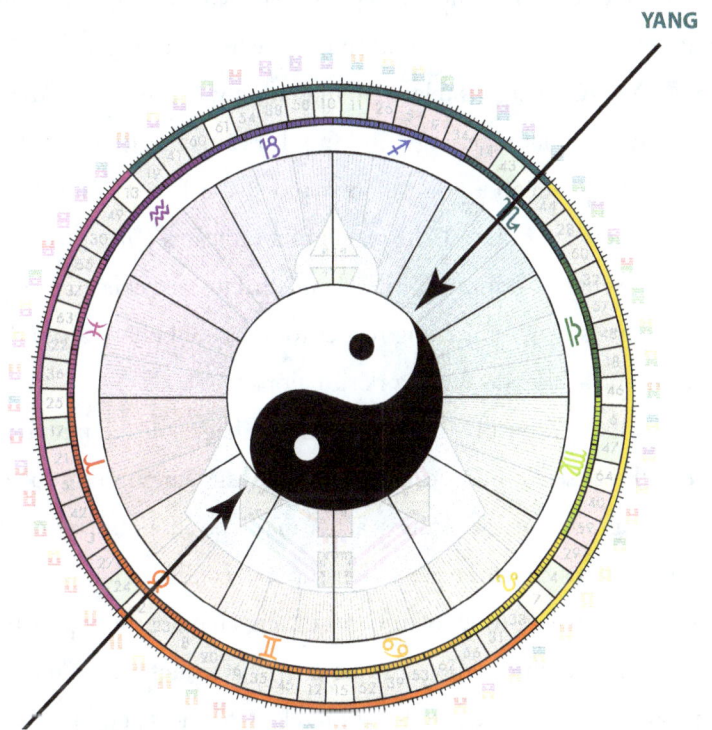

Illustration 12: Mandala Showing Yin/Yang in Wheel
Yin is Receptive (Black) Energy, Yang is Active (White) Energy

YIN Energy represents the feminine, intuition, darkness, coldness, earth, right brain, receptivity, compassion, and submission. It begins in the Second House at the forty-five degree point of Taurus ♉, a fixed Earth Sign, which represents a bulwark for tradition, as well as practicality and stability. It is in the first quadrant of the astrological wheel that relates to inner process and activates inner grounding.

The most Yin Energy in the I-Ching begins at forty-five degrees into the second house. At this location, it has six yin lines in its double trigram hexagram, and is at its most introspective in this position. It is represented by two trigrams called Earth. This double configuration of Earth energy is equivalent to all of the most Yin characteristics possible and acts in its most receptive role. Yin's shy, intuitive energy relates to how you behave, think, and feel when you're alone.

☰ **YANG** Energy is related to masculinity, heaven, warmth, heat, light, action, initiating energy, and fire. It begins in the Eighth House at the forty-five-degree point of Scorpio ♏, a Fixed Water Sign. It is in the third quadrant of the wheel that indicates outer process and interactive energy. It moves the Self outward toward collective expression and transformation. Yang relates to how you behave around others, as in social situations or work situations.

The most Yang Energy in the I-Ching begins at the midpoint of the eighth house. At this location, it has six yang lines in its double trigram hexagram, and is at its most active in this position. It is represented by two trigrams called Heaven. This double configuration of Heaven energy is equivalent to all of the most Yang characteristics possible. It acts as a creative force. Yang's strong energy activates how you behave, think, and feel and influences how you relate to others and to the world.

The lower trigrams of double trigram hexagrams relate to the astrological sign in a chart.

Analyzing the Astrological Wheel

In astrology, there are 12 signs and 12 houses. Each sign has a house it belongs to in which its expression is always strongest. For example, Aries is at home in the First House, Taurus is at home in the Second House

PUTTING TOGETHER THE WHEEL ♦ 41

and so on. Planets too feel most at home in certain signs and when found in their native houses, they are said to "rule" those signs.

Houses and Signs move around the wheel in a counterclockwise direction.

The planetary house and sign positions for any given chart track and show the developmental possibilities for an individual that are uniquely inherent to their life. Chart analysis puts these possibilities in perspective cohesively and in context for the bigger picture and forward momentum in life.

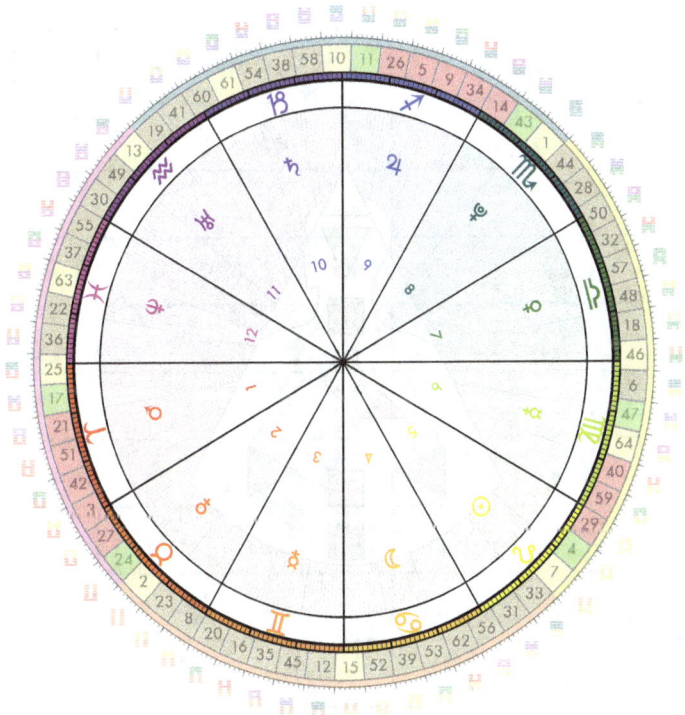

Illustration 15: The Astrology Wheel in the Mandala of Synthesis: An Integration of Science and Intuitive Knowledge

Houses Of The Astrology Wheel And Their Relationship With The Area of Life Cycle Affected

Houses describe the life area impacted by planets and sign. They show contexts of expression and likely situations or circumstances that an individual might experience. Each house has a cusp (or doorway).

The twelve houses are numbered and associated with a particular sign. They're also associated with a planetary ruler, i.e., a planet most at home in that sign.

Sign	Planet	House Number
Aries ♈	Mars ♂	1
Taurus ♉	Venus ♀	2
Gemini ♊	Mercury ☿	3
Cancer ♋	Moon ☽	4
Leo ♌	Sun ☉	5
Virgo ♍	Mercury ☿	6
Libra ♎	Venus ♀	7
Scorpio ♏	Pluto ♇	8
Sagittarius ♐	Jupiter ♃	9
Capricorn ♑	Saturn ♄	10
Aquarius ♒	Uranus ♅	11
Pisces ♓	Neptune ♆	12

Illustration 16: Table of Astrological Sign, Planet, House

PUTTING TOGETHER THE WHEEL ♦ 43

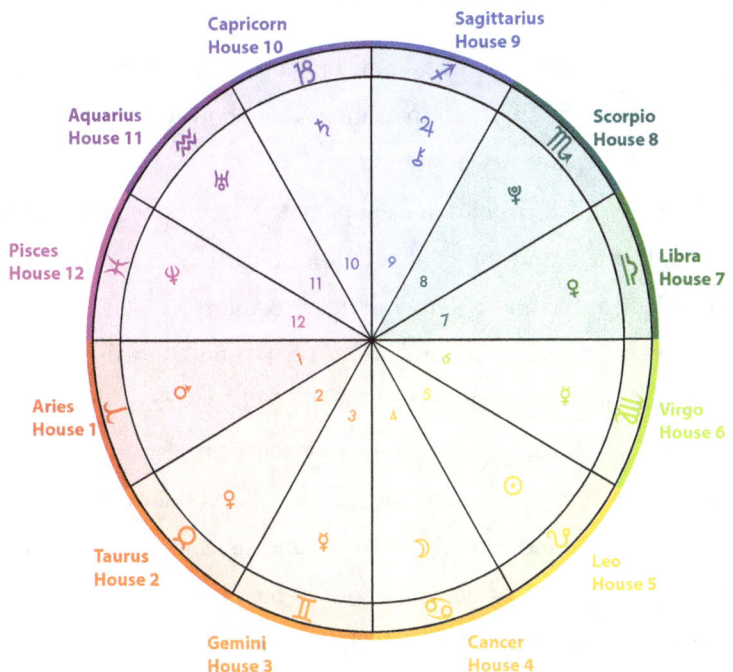

Illustration 17: The Colored Astrological Wheel with Signs, Planets, and Houses

In this image, the planets are indicated by their symbol, and shown in the house with the sign they rule. For example, Venus rules both the second and seventh houses and Mercury rules both the third and the sixth house. Depending on the way the planets land in the chart and how they are configured, the reader must determine how to interpret the sign and its influence on the individual being read.

Since astrological calculations and the construction of the astrological wheel is complex, some of the concepts will be fully explained later in this book. For simplicity of your understanding, I am focusing on the essentials you need to know first and referencing the more complex concepts that will be clearer and more easily understandable later.

What to Look For In a Wheel

Planetary house positions show time cycles. The location of each planet in a specific chart is important as they show where cycles begin as well as their planetary positions.

When studying a chart, look at planetary houses, their quadrants, patterns formed, and the configuration of the planets. You will also want to look at the house signs and their "comfort level", i.e., whether or not a planet falls in its natural house. First, you'll study their position in terms of element, quality, and how they might express themselves.

Next, you'll learn that how relationships express themselves depends on the consciousness of the individual and their choices in life. Then, once you understand the wheel and its components, you will learn about how the wheel is laid out in terms of time and location and how the astrological clock and time are set up in the zodiac wheel. At this point, just notice the terms used in the illustrations such as the Ascendant, Descendant, Midheaven, and IC. Pay attention to these positions as emphasized. The meanings will fall into place later.

The Planetary Signs and their aspects show how they relate. Planetary relationships (aspects) indicate the internal dynamics of the chart, i.e., the ease or tension between them. There are **TRINES** (triangles) and **CONJUNCTIONS** (planets next to each other) that show *ease*, **SQUARES** and **OPPOSITIONS** that show *tension* and other relationships with more subtle nuances.

The Quadrants in the Wheel

The astrological wheel is divided into quadrants that each represents an orientation and perspective approach to life. In order to grasp the intricacies of reading a chart it is important to learn all the components that interact in a chart. The technical components and calculation considerations are presented later in the book because many of those considerations are unnecessary in looking at the body maps used in Noble Sciences' charts. However, learning the nuances of meaning for the houses, signs, planets, and their energy components in the wheel and its quadrants is important in understanding the subtleties of their interactions in a body map.

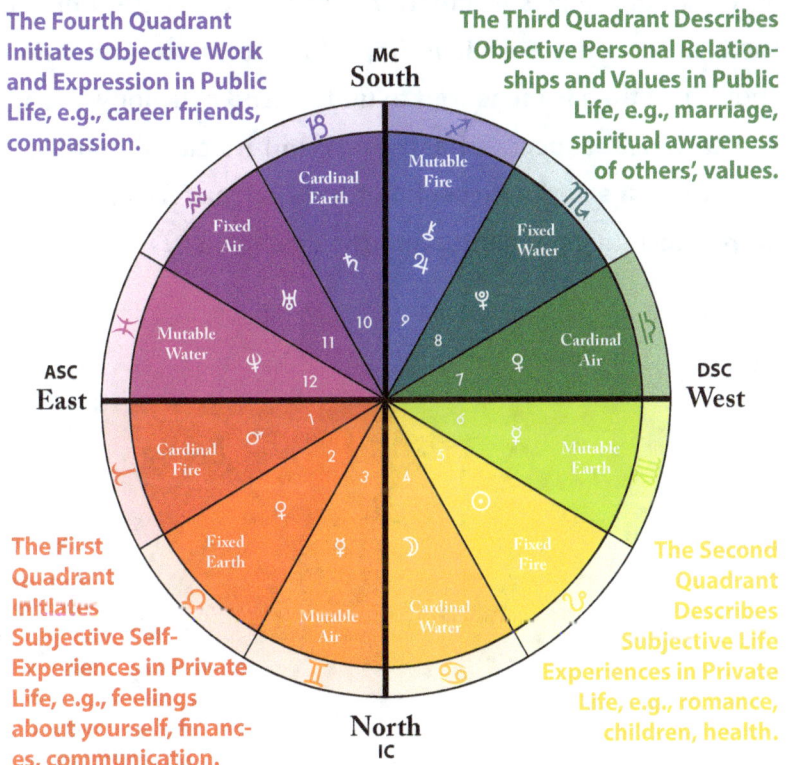

Illustration 18: Astrological Wheel with Quadrants Labeled
http://beyondhumandesign.com/go-beyond-human-design/ancient-wisdom/

I have presented many of the concepts graphically to make understanding easier for you. Once you grasp the basics, you will be prepared to delve more deeply into how charts are calculated and why exact birth data is essential in accurate interpretations.

Note To Reader on The Importance Of Longitude & Latitude: **The zodiac wheel is divided into quadrants that remain constant in meaning based on their position in the wheel regardless of the sign on each house cusp. And, because on Earth, time measurement is based on the motion of the Earth around the Sun, it is important to know something about how time is measured to understand astrology charts. The technical aspects of astrology and its calculations are presented in a separate section for those readers interested in the mechanics of charts and their nuances.**

PART IV

ASTROLOGICAL HOUSES AND SIGN KEYNOTES

The First Quadrant Houses
Initiates Subjective Self-Experiences in Private Life
(Feelings About Yourself, Finances, Communication)

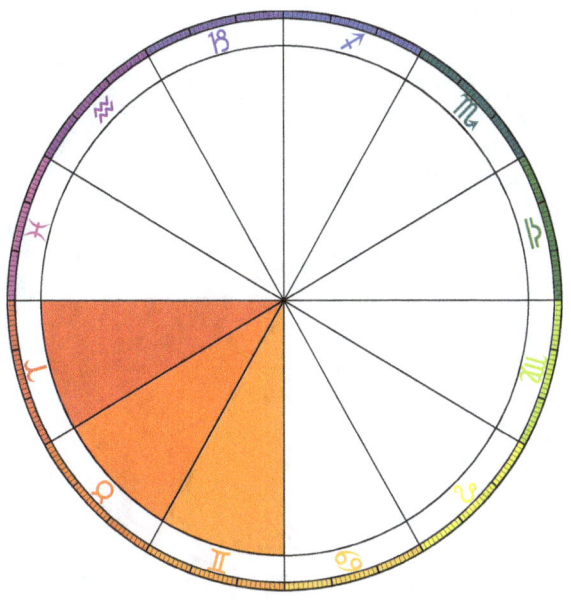

Illustration 19: The First Quadrant Houses

First House: Cardinal/Fire/Aries
Appearance, The Persona/Self, Physical Body, Independent Action

Second House: Fixed/Earth/Taurus
What you 'Own' – Possessions, Things, Values, Attitudes, Self-Esteem

Third House: Mutable/Air/Gemini
Communication, Information, Things Close to you, Writing, Short Term Moves

The Second Quadrant Houses
Initiates Subjective Life Experiences in Private Life (Romance, Children, Health)

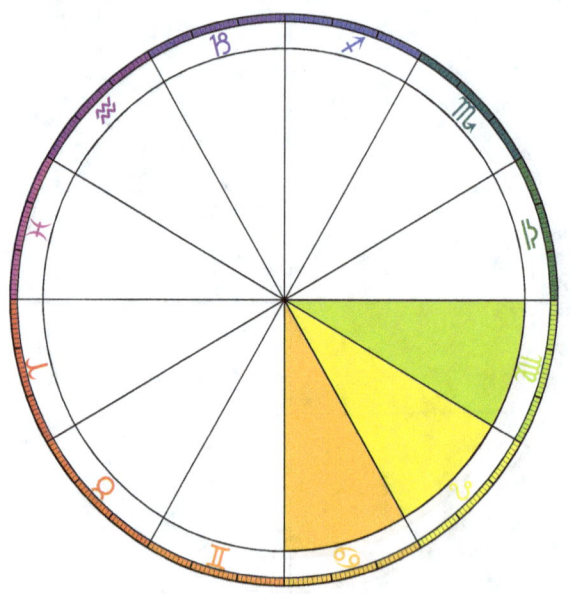

Illustration 20: The Second Quadrant Houses

Fourth House: Cardinal/Water/Cancer
Home and Family, Experiences in Childhood Home, Your Core: What is Important in Your Adult Home

Fifth House: Fixed/Fire/Leo
Creative Self-Expression, Romance, Identity, Children, Drama

Sixth House: Mutable/Earth/Virgo
Health/Healing, Pets, Service, Practicality, Problem Solving

The Third Quadrant Houses
Describes Objective Personal Relationships and Values in Public Life, (Marriage, Spiritual Awareness of Others, Values)

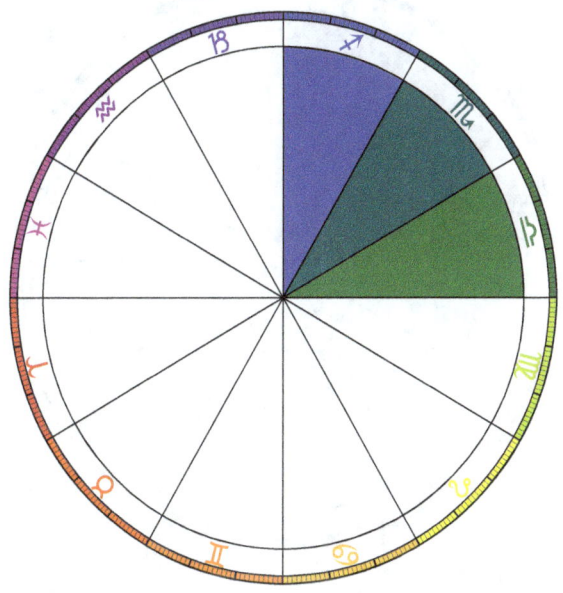

Illustration 21: The Third Quadrant Houses

Seventh House: Cardinal/Air/Libra
Close Relationships, Marriage, Partnerships, Legal Issues

Eighth House: Fixed/Water/Scorpio
Sex, Transformation, Death, Rebirth, The Occult, Politics, What is Private and Hidden, Power

Ninth House: Mutable/Fire/Sagittarius
Philosophy, Higher Education, Truth-Seeking, Publishing, Religion, Beliefs, Superconsciousness, Long Term Moves

The Fourth Quadrant Houses
Describes Objective Work and Expression in Public Life (Career, Friends, Compassion)

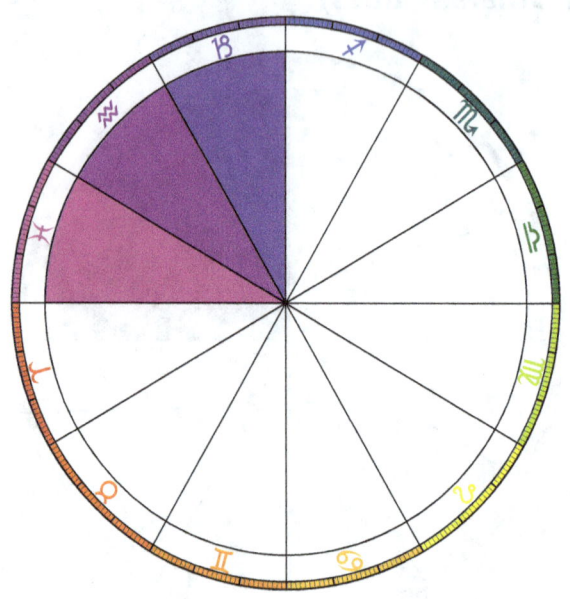

Illustration 22: The Fourth Quadrant Houses

Tenth House: Cardinal/Earth/Capricorn
Social Status, Long-Term Goals, Career, What You Build in Your Life, Authority, Reputation, A Parental Figure

Eleventh House: Fixed/Air/Aquarius

Friendship, The Collective, Hopes and Wishes, The Future, Peers, Groups, Organizations

Twelfth House: Mutable/Water/Pisces

Subconscious, Meditation, Dreams, Illusions, Universal Love, What Is Behind the Veil, Isolation, Fantasy

ASTROLOGICAL HOUSES AND SIGN KEYNOTES ♦ 53

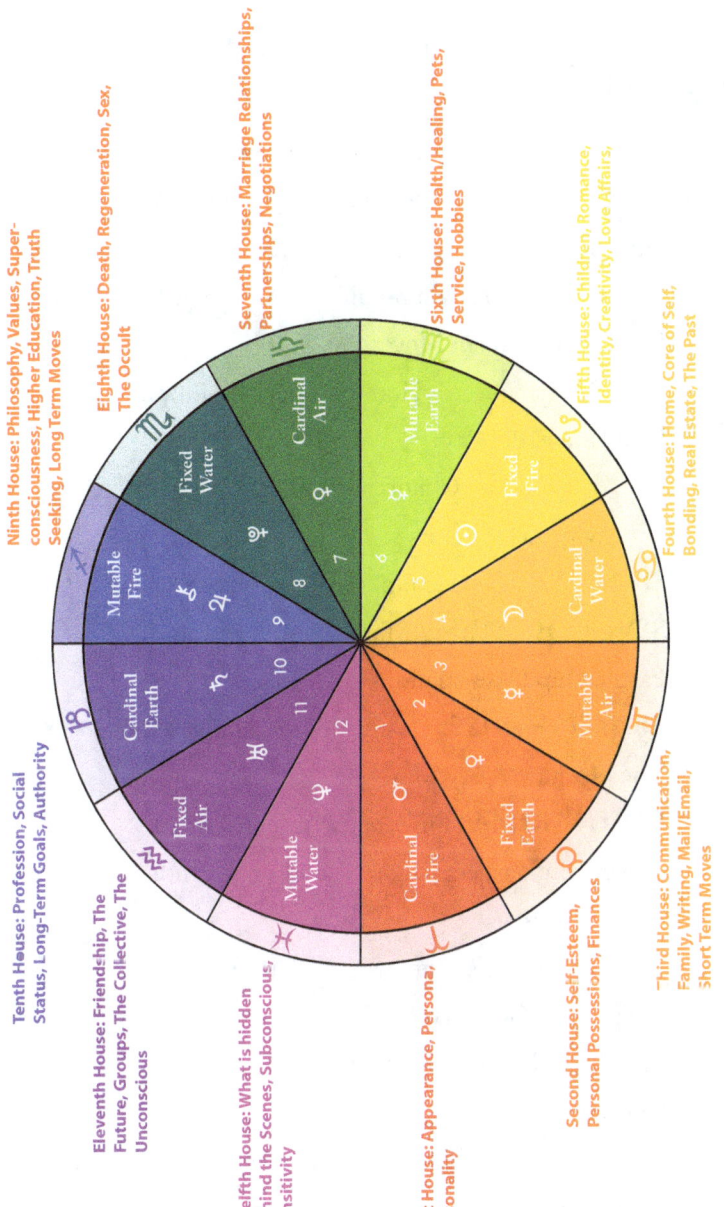

Illustration 23: Astrological House Keynotes and Areas of Life Affected

The following are descriptions of each house:

- **FIRST HOUSE:** Appearance, Persona, Personality
- **SECOND HOUSE:** Self-Esteem, Personal Possessions, Finances
- **THIRD HOUSE:** Communication, Family, Writing, Mail/Email, Short-Term Moves
- **FOURTH HOUSE:** Home, Core of Self, Bonding, Real Estate, the Past
- **FIFTH HOUSE:** Children, Romance, Identity, Creativity, Love Affairs
- **SIXTH HOUSE:** Health/Healing, Pets, Service, Hobbies
- **SEVENTH HOUSE:** Marriage Relationships, Partnerships, Negotiations
- **EIGHTH HOUSE:** Death, Regeneration, Sex, The Occult
- **NINTH HOUSE:** Philosophy, Values, Superconsciousness, Higher Education, Truth Seeking, Long-Term Moves
- **TENTH HOUSE:** Profession, Social Status, Long-Term Goals, Authority
- **ELEVENTH HOUSE:** Friendship, The Future, Groups, the Collective
- **TWELFTH HOUSE:** What is Hidden Behind the Scenes, Subconscious

Planets in the Astrology Wheel: Their Areas of Experience and Behavior

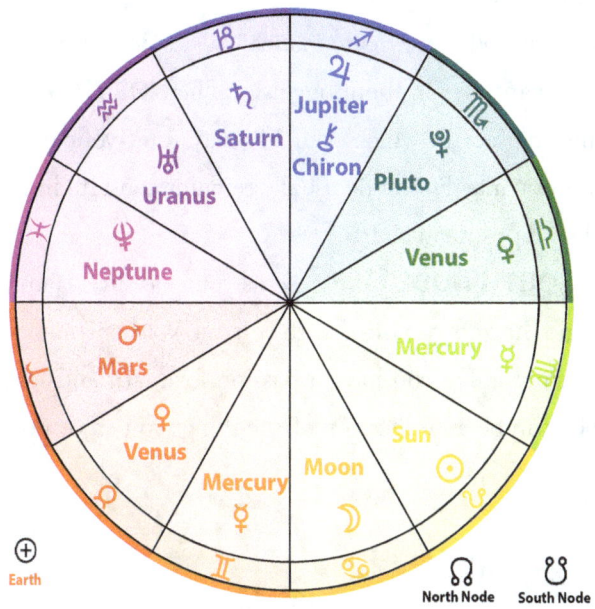

Planets Describe Characteristic ways of: Acting, Thinking, Communicating

Illustration 24: Planets in the Wheel in the Houses They Rule

Astrological signs categorized by their qualities, elements, and characteristics describe the energy and the kinds of influence the sign has on an individual, event, or relationship. Astrological houses describe the area the sign or planetary body impacts.

Planets describe characteristic ways of:
- Acting.
- Thinking.
- Communicating.

There are twelve primary planetary bodies generally considered in astrological work, along with the North and South Nodes: the **Sun**, the **Earth** (always 180 degrees opposite the Sun), the **Moon, Mercury,**

Venus, **Mars**, **Jupiter**, **Saturn**, **Uranus**, **Neptune**, **Pluto**, and **Chiron,** considered an asteroid but one of great significance in astrological work.

Although nodes are mathematical calculations, not planets, they reveal important details about our lives. The **NORTH NODE** ☊ represents the manner in which your life plays out, your life's purpose, destiny, and karma. This node usually remains constant in your prenatal, natal, and postnatal charts.

The **SOUTH NODE** ☋ represents your past lives, familiarity, belonging, and the gifts you bring with you into your current life. Are you a violin virtuoso? Do you have a passion for math and science? The south node can be an indication of the things you enjoy doing.

ASTROLOGICAL HOUSES AND SIGN KEYNOTES ♦ 57

Planets and Keywords for Reading Them	
♂	Mars – Action, Conflict, Courage, Assertiveness, Initiative
♀	Venus – Harmony, Beauty, Affection, Relating, Balance
☿	Mercury – Communication, The Mind, Connection, Spirit and Matter, "Quicksilver"
☽	Moon – Emotional Feelings, Attachments, Children, Instincts, Feminine Receptivity, Nurturing
☉	Sun – Vitality, Individuality, Spirit, Will, Energy
☿	Mercury – Communication, The Mind, Connection, Spirit and Matter, "Quicksilver"
♀	Venus – Harmony, Beauty, Affection, Relating, Balance
♇	Pluto – Deep Transformation, Regeneration, Resources of Others
♃	Jupiter – Good Fortune, Expansion, Growth, Optimism, Positive, Consciousness, Awareness
♄	Saturn – Limitation, Discipline, Effort, Authority, Life Challenges
♅	Uranus – Originality, Independence, Unexpected, Uniqueness
♆	Neptune – Compassion, Cloudy, Confusion, Sensitivity, Imagination
⚷	Chiron – Healing, Sensitive, Perceptive, Expansive, Conscious Focus of Awareness
☋☊	The South Node and North Node are planetary points related to your life purpose and destiny.
⊕	The Earth is a planetary point related to your grounding.

Illustration 25: Astrology Planet Keynotes: Experience & Behavior

Signs in the Astrology Wheel: Ways of Functioning

Signs describe characteristics associated with the expression of:
- The Element – Fire, Earth, Air, Water
- The Quality (Mode) – Cardinal, Fixed, Mutable
- The House they are in and rule – 12 astrological houses each ruled by a sign and planet

The Twelve Signs are:
- **ARIES** ♈: Activating, Willful, Energizing, Smart
- **TAURUS** ♉: Practical, Firm, Dependable, Stubborn
- **GEMINI** ♊: Duality, Communications, Changeable, Friendly, Cheerful
- **CANCER** ♋: Nurturing, Sensitive, Intuitive, Emotional
- **LEO** ♌: Vitality, Strong, Energetic, Dramatic
- **VIRGO** ♍: Detailed, Rational, Methodical, Practical
- **LIBRA** ♎: Peace Makers, Interactive, Related, Balance
- **SCORPIO** ♏: Intense, Loyal, Sexy, Transformative, Blunt
- **SAGITTARIUS** ♐: Straightforward, Philosophical, Truth-Seeking, Optimistic
- **CAPRICORN** ♑: Structured, Disciplined, Ambitious, Responsible
- **AQUARIUS** ♒: Unique, Collective, Revolutionary, Independent
- **PISCES** ♓: Mystical, Private Reflective, Sensitive

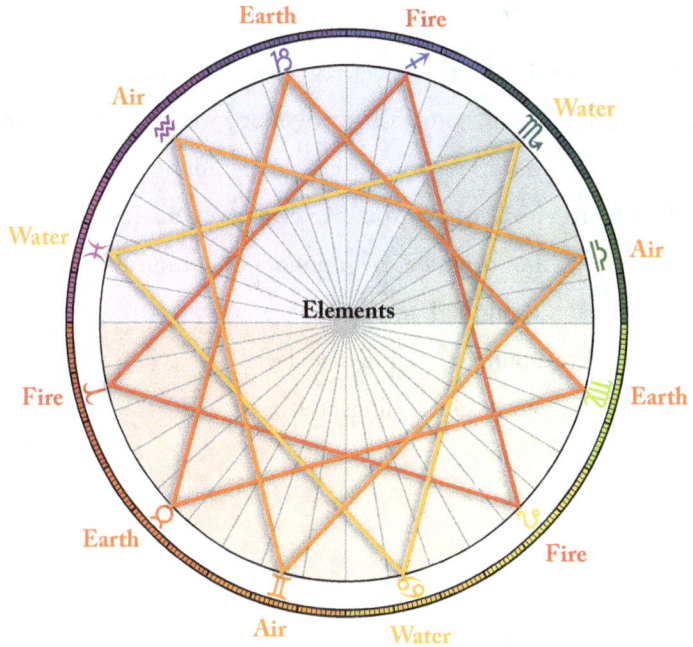

Illustration 26: Trines of Elements

The Four Astrological Elements

Elements describe the basic nature or temperament of a sign. The four elements are Fire, Earth, Air, and Water. Each has a sign associated with it and repeats three times in the wheel (4 elements, 3 times = 12 houses).

Signs in the same element are 120 degrees (120°) apart. The same elements form a triangle, e.g., fire and fire, earth and earth, etc.

Triangles & Trines △

Trines represent an ease of energy flow. They also show that personality, character, or life processes are working well in life as the planets and archetypes are working together in cooperation.

The four elements and their characteristics:
- **FIRE:** action-oriented, self-confidence, enthusiasm
- **EARTH:** grounded, methodical, practical
- **AIR:** communication, intelligence
- **WATER:** intuition, feelings, instinct

The Three Signs in the Element FIRE

Creative, Initiating, Self-Assured, Expansive, Enthusiastic, Impulsive

The Fire Elements create triangles with each other.

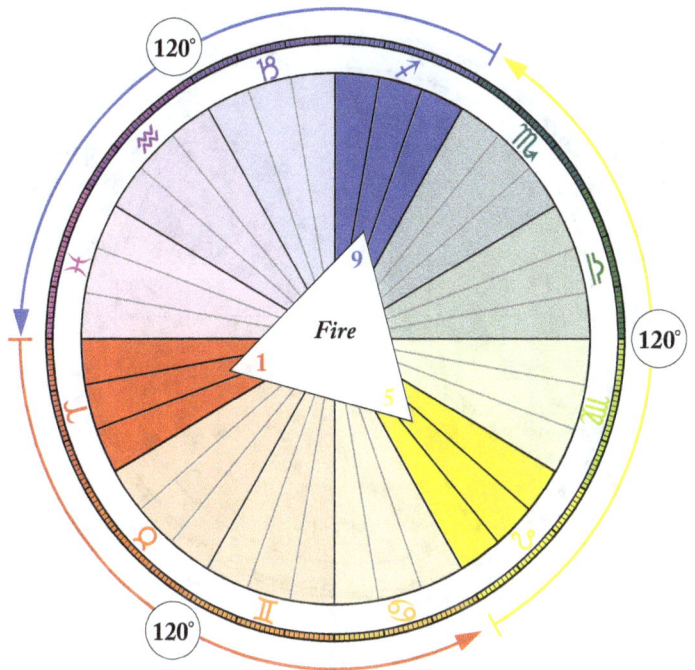

Illustration 27: The Fire Elements

The Fire Elements are:

- **ARIES** ♈: Active, Pioneering
  - Ruled by Mars ♂
  - Rules the First House
- **LEO** ♌ : Strong, Energetic
  - Ruled by the Sun ☉
  - Rules the Fifth House
- **SAGITTARIUS** ♐: Inspired, Truth-Seeking
  - Ruled by Jupiter ♃ and Chiron ⚷
  - Rules the Ninth House

The Three Signs in the Element Earth

Grounded, Caring, Methodical, Practical, Realistic, Serious

The Earth Elements create triangles with each other.

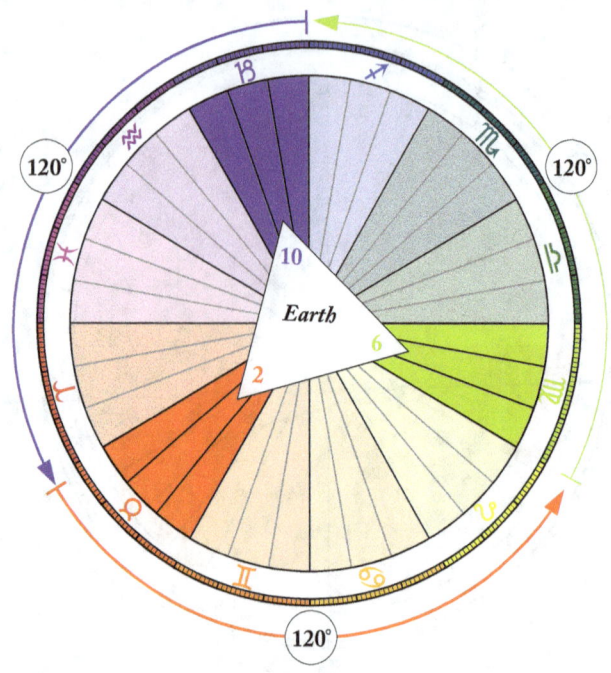

Illustration 28: The Earth Elements

The Earth Elements are:
- **TAURUS** ♉: Dependable, Fixed
  - Ruled by Venus ♀
  - Rules the Second House
- **VIRGO** ♍: Perfectionist, Health Oriented
  - Ruled by Mercury ☿
  - Rules the Sixth House
- **CAPRICORN** ♑: Responsible/Structured
  - Ruled by Saturn ♄
  - Rules the Tenth House

The Three Signs in the Element Air

The Mind, Ideas, Awareness, Communication, Exchange of Information, Interaction

The Air Elements create triangles with each other.

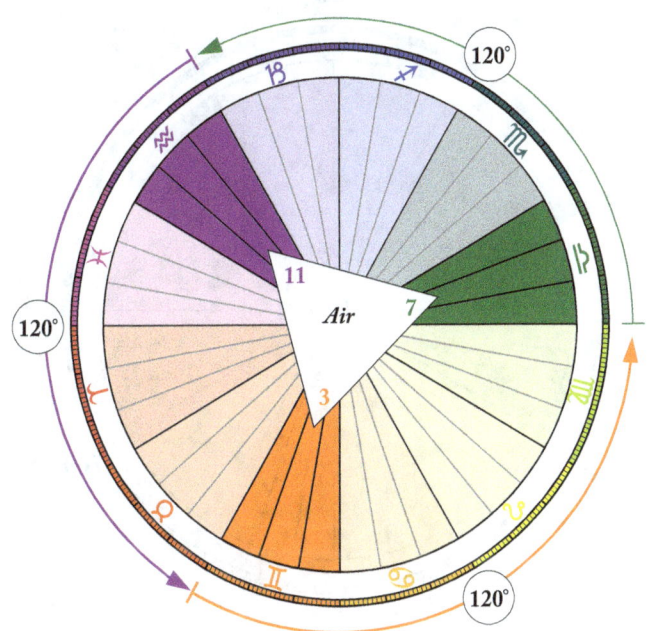

Illustration 29: The Air Elements

The Air Elements are:
- **GEMINI** ♊: Changeable, Communicative
  - Ruled by Mercury ☿
  - Rules the Third House
- **LIBRA** ♎: Balance, Peace-Seeking
  - Ruled by Venus ♀
  - Rules the Seventh House
- **AQUARIUS** ♒: Innovation, Independent
  - Ruled by Uranus ♅
  - Rules the Eleventh House

The Three Signs in the Element Water

Sensitive, Emotional, Empathic, Intuitive, Receptive, Unconscious

The Water Elements create triangles with each other.

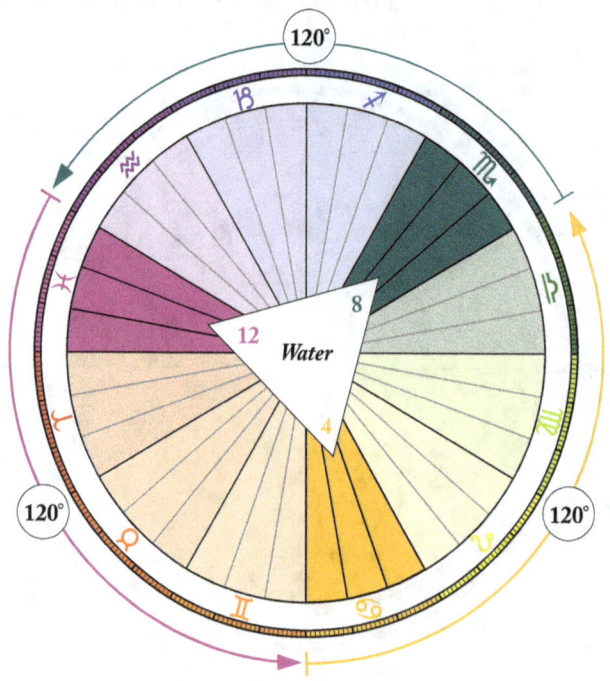

Illustration 30: The Water Elements

The Water Elements are:

- **CANCER** ♋: Nurturing, Home-Oriented
  - Ruled by Moon ☽
  - Rules the Fourth House
- **SCORPIO** ♏: Intensity, Transformative
  - Ruled by Pluto ♇
  - Rules the Eighth House
- **PISCES** ♓: Sensitive, Compassionate
  - Ruled by Neptune ♆
  - Rules the Twelfth House

Trines △ on Cusps and House Midpoints

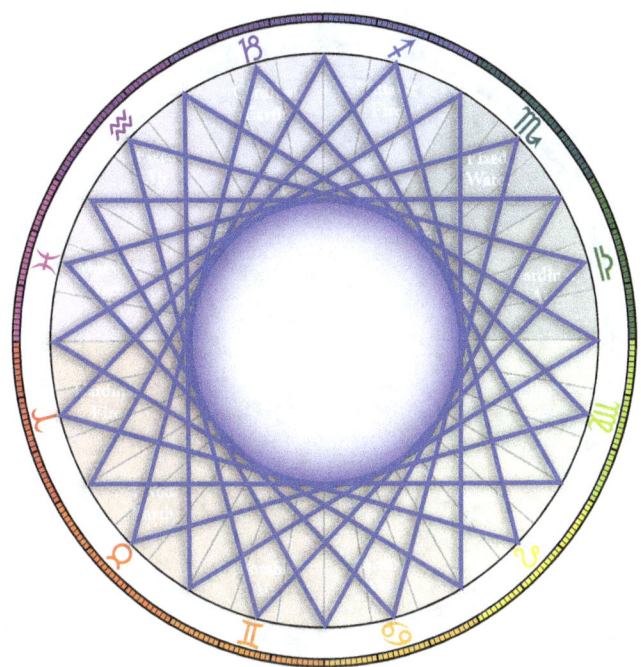

Illustration 31: Mandala Cusps and Midpoint Trines

Astrological Elements are used to describe the basic nature or temperament of a sign, which are always 120 degrees apart. 120° forms a TRINE aspect.

Trines are generally facilitative aspects that smooth and balance.

Midpoints can fall between any two cusps, planets, or signs. They can take place every 15 to 60 degrees. Any time there are two points on a wheel, there's a midpoint. The interpretation of midpoints depend on calculated degree combinations and permutations of midpoints of the degrees of the related planets and/or cusps under consideration.

How to Read Triangles △ (Trines)

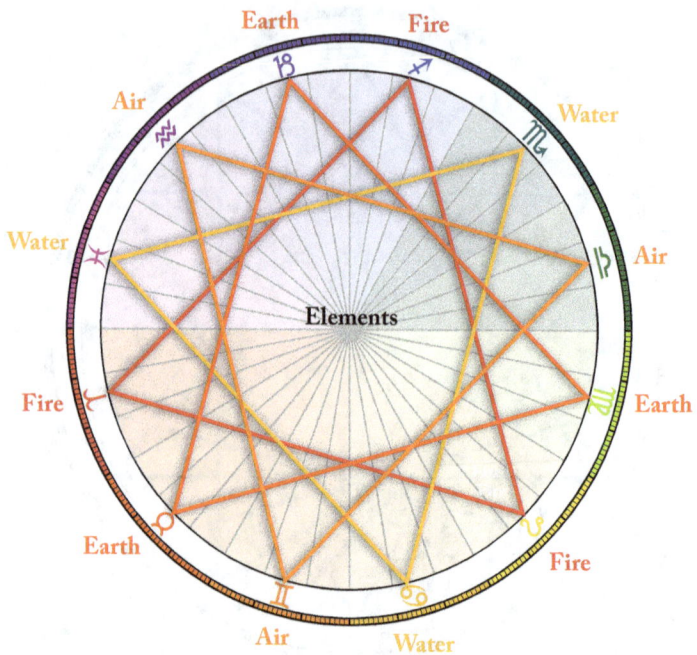

Illustration 32: Element Triangles

The 120° difference between two planets makes it trine even if the third side of the triangle is absent.

The elements, their planets, and houses indicate the areas of life concerned and how the time operates from its three phases: beginning, middle, and end. When two or three Elements relate in the wheel, forming Trines:

- **FIRE** to **FIRE**
- **EARTH** to **EARTH**
- **AIR** to **AIR**
- **WATER** to **WATER**

The Three Astrological Qualities (Modes) Squares □ & Oppositions ☍.
Astrological Qualities describe the nature of a Sign's expression and its' natural house position.

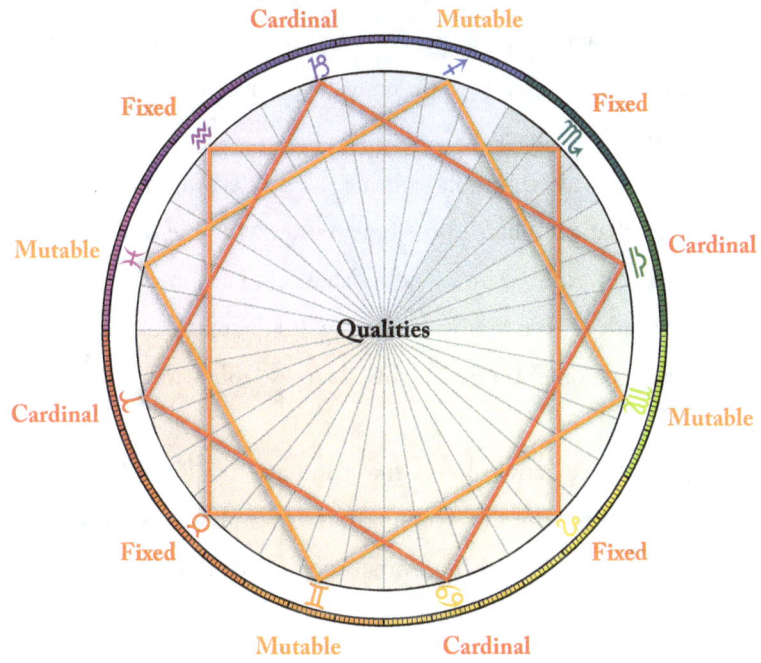

Illustration 33: Qualities: Squares

There are Three Qualities: Cardinal, Fixed, and Mutable

Each quality has a house associated with it in a quadrant and repeats four times in the wheel (3 Qualities, 4 times = 12 houses).

The three Qualities are:

- **CARDINAL:** Since Cardinal signs are the pivotal locations of starting points within the astrological wheel (ASC, MC,

DESC, & IC), they are associated with being initiators. People born under Cardinal signs have a tendency to be trailblazers, visionaries, and leaders, although they don't always finish what they start.

- **FIXED:** Fixed signs are related to stability and loyalty, and are known for being steadfast and resolute. People born under Fixed signs often take orders well and usually see a project through to the end.
- **MUTABLE:** Mutable signs are associated with adaptability and changeability. Those born under Mutable signs are often flexible, experimental, inconsistent, and accommodating.

Signs with a given quality are always 90 degrees (90°) apart, forming a square, or one hundred eighty degrees (180°) apart, forming an opposition.

Squares have active energy that can have a pulling effect on you. This creates a pressure that you might resist. Sometimes this energy is supported by other factors and sometimes it isn't, but it is always moving and acting within our chart.

The Four Signs with Cardinal Quality (Mode)

The Cardinal angles of the wheel lead the quadrant and begin things. The angles show actions/characteristics that are goal-oriented, purposeful, ambitious, self-starting, driven, and enthusiastic.

Cardinal quality (mode) houses and signs create squares with each other.

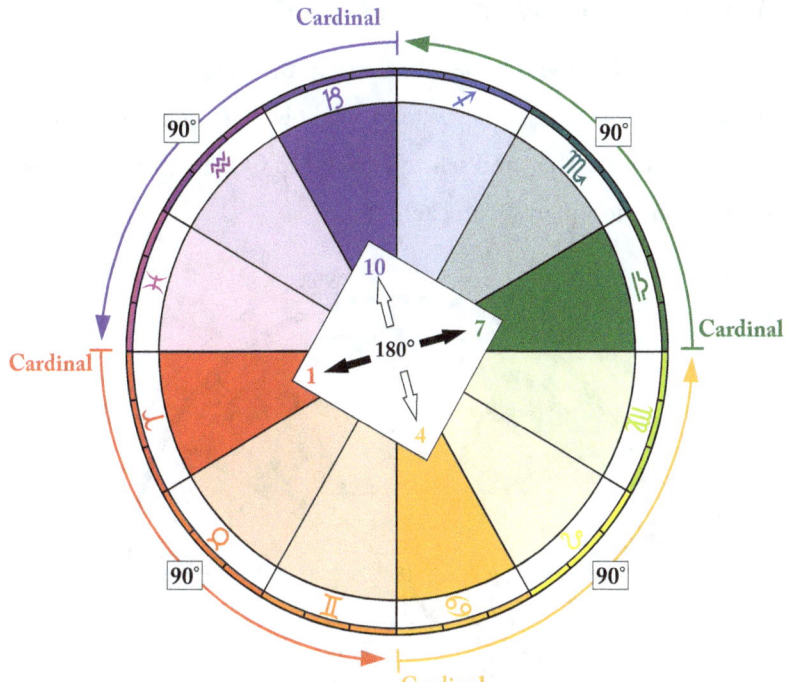

Illustration 34: The Cardinal Squares and Oppositions

The Cardinal Houses are:

- **ARIES ♈**: Fire, Active
  - Ruled by Mars ♂
  - Rules the First House
- **CANCER ♋**: Water, Emotion
  - Ruled by Moon ☽
  - Rules the Fourth House
- **LIBRA ♎**: Air, Balance
  - Ruled by Venus ♀
  - Rules the Seventh House
- **CAPRICORN ♑**: Earth, Lessons/Limitations
  - Ruled by Saturn ♄
  - Rules the Tenth House

The Four Signs with Fixed Quality (Mode)

The Fixed houses of the wheel are in the middle of each quadrant and anchor process. They tend to show characteristics that are patient, steady, reliable, and self-reliant. Fixed Quality (Mode) Houses and Signs create squares with each other.

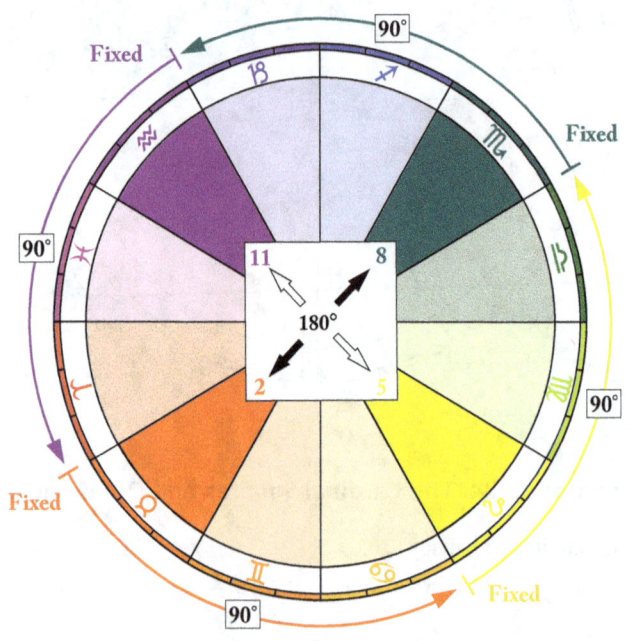

Illustration 35: The Fixed Squares and Oppositions

The Fixed Houses are:
- **TAURUS** ♉: Earth, Balance
  - Ruled by Venus ♀
  - Rules the Second House
- **LEO** ♌ : Fire, Initiating
  - Ruled by the Sun ☉
  - Rules the Fifth House

ASTROLOGICAL HOUSES AND SIGN KEYNOTES ♦ 71

- **SCORPIO** ♏: Water, Transformation
  - Ruled by Pluto ♇
  - Rules the Eighth House
- **AQUARIUS** ♒ : Air, Groups
  - Ruled by Uranus ♅
  - Rules the Eleventh House

The Four Signs with Mutable Quality (Mode)

The Mutable houses of the wheel are the last house in each quadrant. They show characteristics that are flexible, adaptable, resourceful, and versatile.

The Mutable Quality (Mode) Houses and Signs create squares with each other.

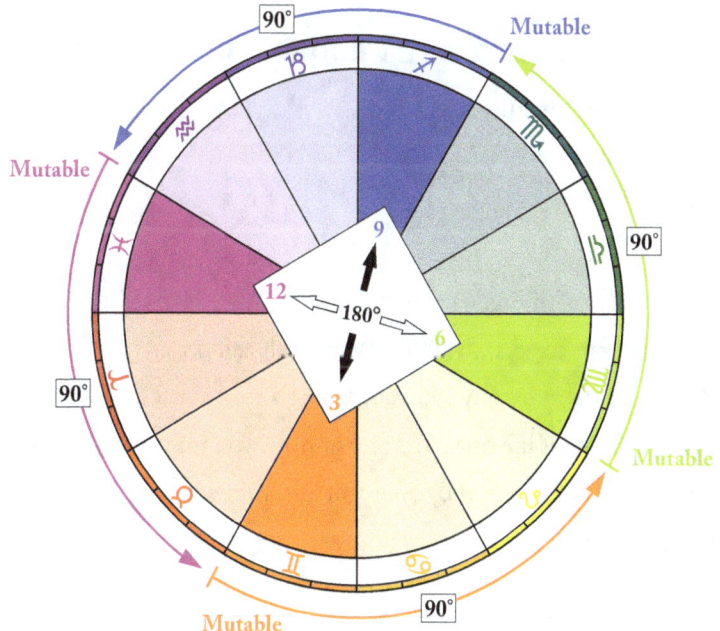

Illustration 36: The Mutable Squares and Oppositions

The Mutable Houses are:

- **GEMINI** ♊ : Air, Duality
  - Ruled by Mercury ☿
  - Rules the Third House
- **VIRGO** ♍: Earth, Work
  - Ruled by Mercury ☿
  - Rules the Sixth House
- **SAGITTARIUS** ♐: Fire, Expansive
  - Ruled by Jupiter ♃ and Chiron ⚷
  - Rules the Ninth House
- **PISCES** ♓: Water, Compassion
  - Ruled by Neptune ♆
  - Rules the Twelfth House

How to Read Squares □ and Oppositions ☍

The three Qualities are:

- Cardinal
- Fixed
- Mutable

The signs with a given quality are always 90° or 180° apart. 90° forms a Square aspect. 180° forms an opposition. Squares and opposition aspects generally activate friction and tension. Oppositions represent a conflict in your life. For example, you may feel a tug-of-war relating to the pain caused by an event and may seesaw between anger and forgiveness.

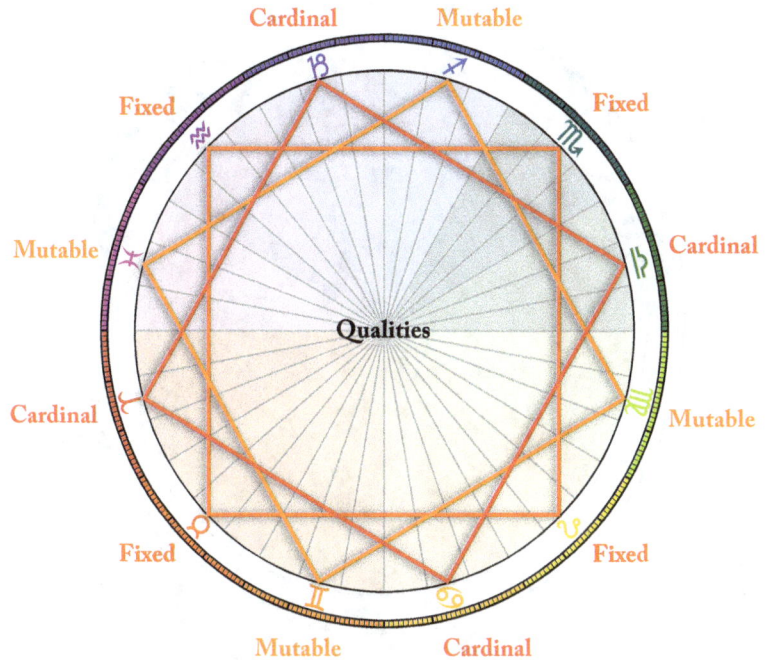

Illustration 37: Qualities: Squares, Cardinal, Fixed, Mutable

Midpoint on Cusps and House

Every zodiac position makes an exact square (90°) with every other like-quality zodiac position of the same zodiac degree, e.g., 17°30'15" ♊, (Gemini, an Air sign), exactly squares 17°30'15" ♍, (Virgo, an Earth sign) and exactly opposes 17°30'15" ♐, (Sagittarius, a Fire sign).

The qualities, their planets, and houses indicate the areas of life concerned and how the process operates from its three phases: beginning, middle, and end.

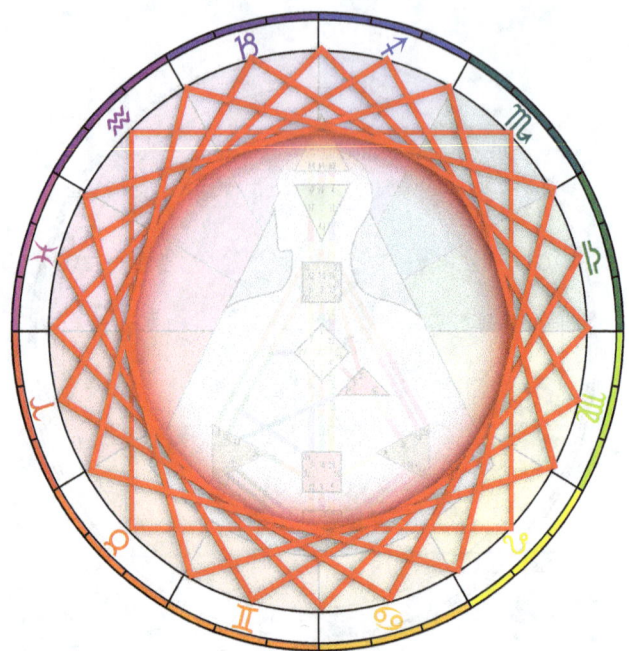

Illustration 38: Mandala Cusps and Midpoint Squares

When Qualities appear in similar degrees in the wheel they are in Square or in Opposition if they are:

- Cardinal to Cardinal
- Fixed to Fixed
- Mutable to Mutable

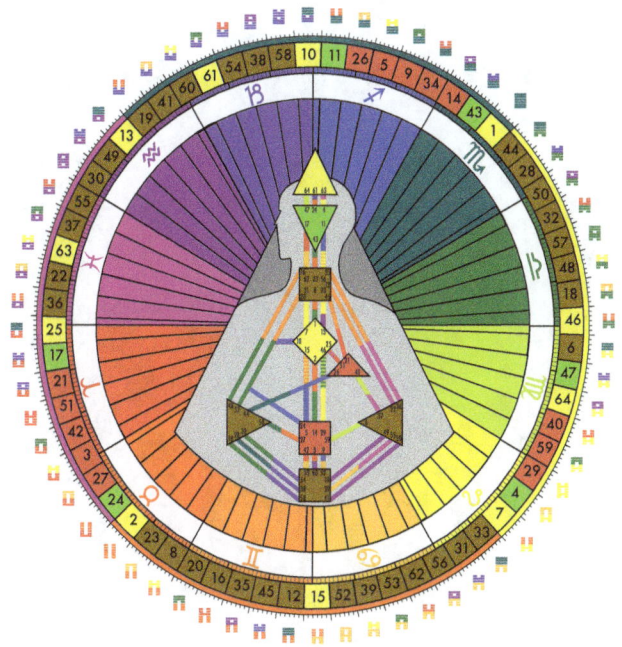

Part V

ASTROLOGICAL CALCULATIONS: THE IMPORTANCE OF ACCURACY

Understanding the Mechanics of Astrology and the Charts

ASTROLOGICAL CALCULATIONS ♦ 77

The Clock in the Wheel: Counting Time and Houses

The astrological wheel is divided into 24 hours, equivalent to one day on Earth. **TIME** is measured in a **CLOCKWISE** direction, just like a regular clock. The division of each house is composed of 2-hour intervals.

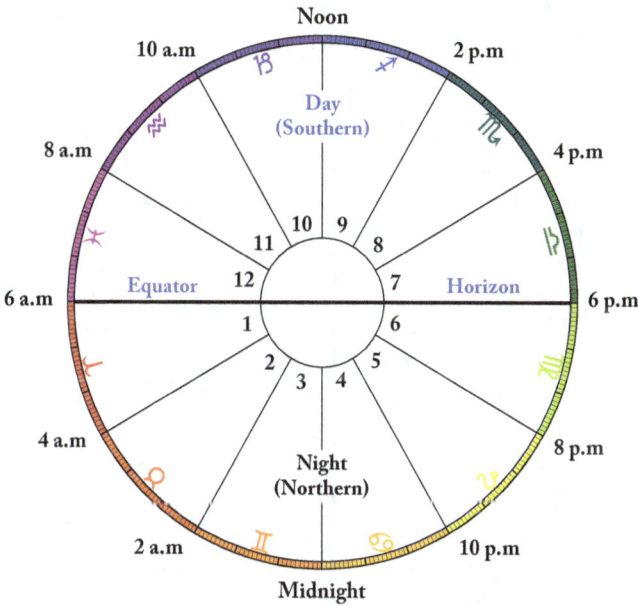

Illustration 39: The Astrological Clock

The Astrological Clock is different from a regular clock. In astrology, noon is at the twelve o'clock position, 6PM is at the 3 o'clock position, and midnight is at the six o'clock position of a regular time clock.

The first house counted on the astrological chart is Aries, the House of the Rising Sun, which is located at 6AM GMT (Greenwich

Mean Time or the Prime Meridian) on the astrological wheel. The Ascendant (Rising Sign) is always found at this position. From Aries, the houses continue to be counted in a counterclockwise direction—Aries, Taurus, Gemini, Cancer, and so on.

The position of House 1, House 2, and so on though House 12 are always placed in the same position in a chart. House 1 is always on the East of the astrological wheel, (9 o'clock in a time clock) and its Sign is determined by the calculated Ascendant (Rising Sign). In other words, the Ascendant always falls in the first house.

Hemispheres Divide the Wheel
East – North – West – South

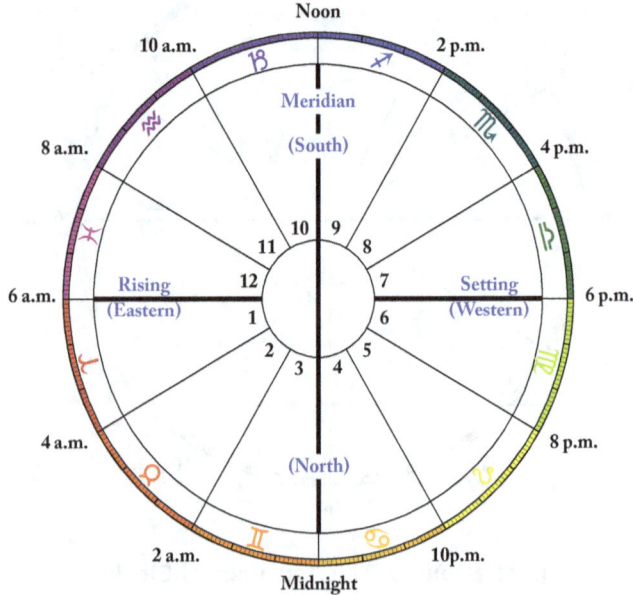

Illustration 40: East-North-West-South in the Astrological Wheel

Hemispheres, Time, and How to Read the Wheel

HOUSE SIGNS move in a **COUNTERCLOCKWISE** direction. The astrological houses begin counting on the horizon at 6 am, moving counterclockwise. Each house division counts two hours. The houses in their calculated signs trace internal development as it unfolds prior to expressing itself in the outer world. The houses are always numbered from one to twelve regardless of where the calculated signs and planets are in an individual's chart. Thus, although an event is oriented in terms of the perspective of the event's time and location, natural counting of the houses remains the same. Each house has a specific meaning as determined by the natural zodiac wheel that calculates the East or rising position at six o'clock in the morning. Individual chart calculations may place signs in different houses than the natural zodiac wheel but the flavor of the natural house position remains constant in the charts.

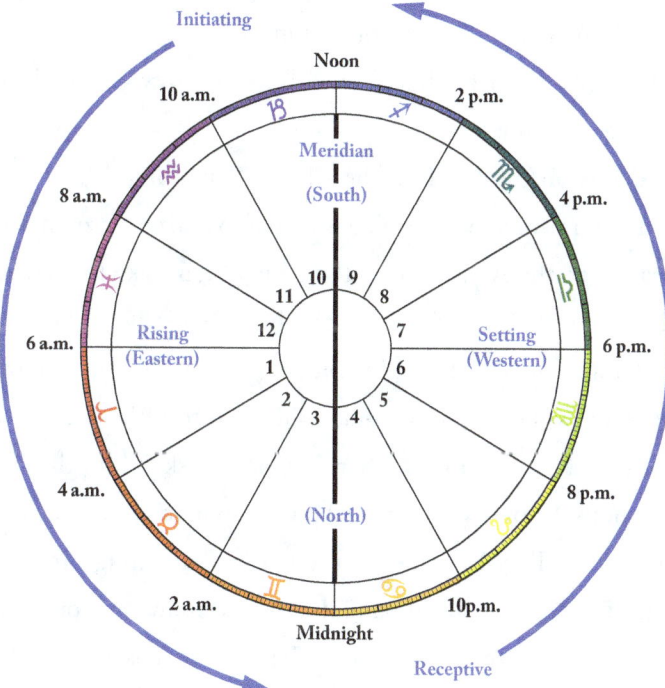

Illustration 41: Telling Time in the Astrological Clock

The meaning of the houses, the planets, and the signs are influenced by their quadrant and house position in the wheel.

If you imagine that you're standing directly in the center of your own astrological chart, you'll see that the hemisphere directions in Astrology are different from the directions on a map. If you're facing south, it's at the top of the chart (in front of you), North is at the bottom (behind your head), East is on your left side, and West is on your Right.

Think of it this way: while you were in the womb, your mother was sitting in a city, town, village, or building somewhere on this planet, and all the while, the stars up in the heavens moved in various configurations over her head in that place and time. Your astrological chart is like a perfect snapshot of the stars and planets above you taken from the point on Earth from which you first entered it at that moment. You are at the center of that image.

On an astrological chart, time and space travel in a clockwise direction, as explained below:

EAST is on the **LEFT** in the wheel. Think of this as the place where the sun rises every morning, which also makes sense since this is the location of the Ascendant or the Rising Sign. Like the dawn of a new day, this is the point for the beginning of your life and is equated with initiating energy. The Rising Sign represents the way you're seen by the world, which could be quite different from the way in which you see yourself, as it uses the persona as a mask. Thus, the sign on the first house shows something about your manifesting personality characteristics. The Ascendant represents the initiating presentation of yourself, or of the issue at hand; it is how you present your self and your personality. In the natural wheel the Ascendant is 6 am GMT, the transition time between night and day or sleep and waking.

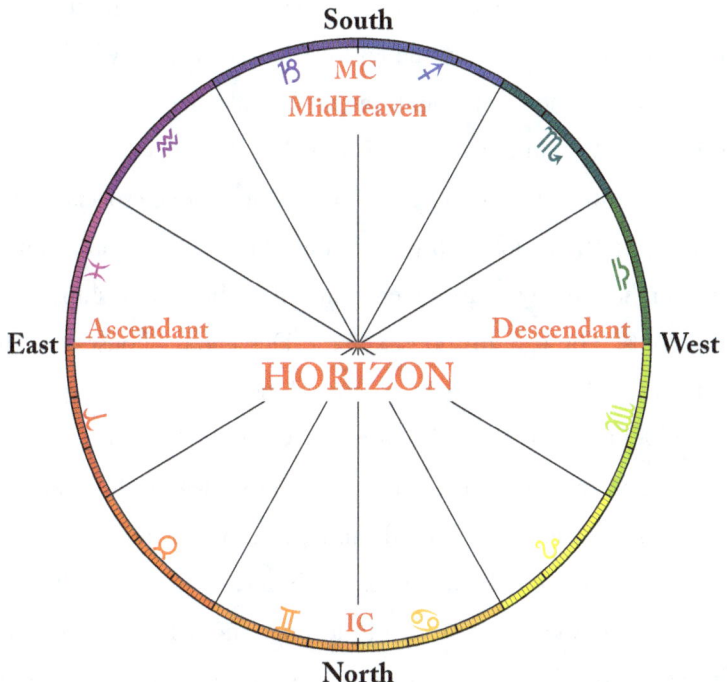

Illustration 42: Time and Space in the Astrological Wheel

SOUTH is at the **TOP** of the wheel. A good way to remember this is to think of it as high noon—the point where the sun would be brightest in the sky, shining down on you as you go about your day. This is where the MC (Medium Coeli) or Midheaven, or the 10th house cusp, is located. This position represents your ambitions, your ability to have the career you want, and your status within society. Some astrologers describe it as how you present yourself to the public. Noon is the time of day when you are likely at the height of wakefulness and when you are interacting with other people. Thus, it indicates how you express yourself in the world.

WEST is on the **RIGHT** in the wheel. This can be thought of as the point where sun sets at the end of the day. It is a twilight space, where day transitions into night. In the natural wheel it corresponds to 6 pm GMT. Within this location is the Descendant, which represents the manner in which you express yourself to others, especially in relationship to those closest to you. As the sun sets, your opportunity to reflect on how you want to present yourself in the relationships in your life unfolds. The seventh house, or the Descendant house, describes legal relationships and intimate partnerships.

NORTH is at the **BOTTOM** of the wheel and can be thought of as midnight, or the place where the moon shines brightest at night. This is the location of the IC (Imum Coeli) or Lower Heaven. This point represents your foundations in life, including the psychological causes for the way you express your feelings and behavior. This fourth house describes the core components of your self. As the IC represents the beginning of your life and foundational roots (and family), it can also represent the end. In the wheel it is the house that represents a private sector of the chart, most likely a time when sleep and dreaming occurs, and thus, it shows a connection with your deepest self.

Above The Horizon (Medium Coeli or MC)

The upper half of the wheel is home to the **MC**, or **MIDHEAVEN**, which is in the **SOUTHERN SECTOR** of the wheel, also the location of high noon. This space shows objective life areas primarily concerned with public, practical, and "visible" things, e.g., career, and friends. The MC shows at its center the core of what and how you present to the world. This represents your goals, dreams, achievements in the eyes of the world, and your reputation in society.

HOUSE TEN, where the **MEDIUM COELI (MC)** is located, is always in the same position in the wheel regardless of its Cusp Sign.

A "cusp sign" is the point at which one sign transitions into another. For example, if you were born, sometime between June 17th to the 23rd, you were born at the cusp of Gemini and Cancer, indicating that you might be intellectual as well as overly emotional.

Consider the following illustration:

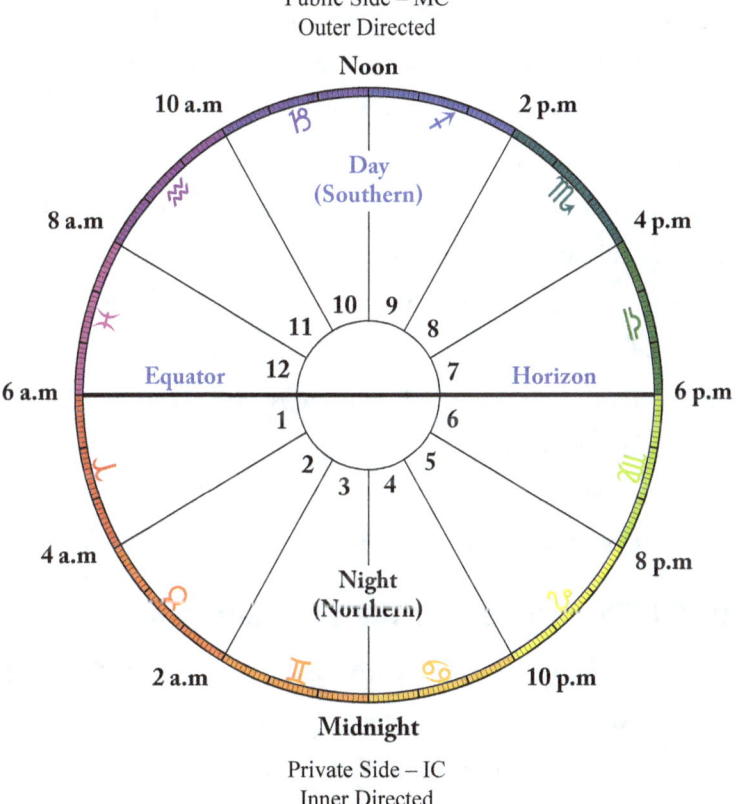

Illustration 43: Directions, Time, and Orientation

Below The Horizon (Imum Coeli or IC)

The area below the horizon, the **NORTHERN HEMISPHERE** of the wheel, relates to our subjective life and is primarily concerned with private matters—things invisible to "outsiders," e.g., self-esteem, self-talk. It is the location of the **IC** and shows what is at the core or center of your private self. It represents the earliest foundations of our lives, as well as our parents, who were responsible for nurturing us. It is here within us that our ancestral legacy lurks, waiting to be unlocked so that we can claim it and move forward to achieve our true destiny.

You can think of this as the location for the kinds of thoughts that go on in your head, the conversation you have with yourself in your mind. Once again, if you were standing in the center of your own chart, this area would represent the place that would be behind your head. The IC is in the fourth house when it is **MIDNIGHT** on the astrological clock. This is the private sector of the chart. It indicates a connection with the deepest self, which most likely occurs during sleep when you are dreaming.

HOUSE FOUR, where the **IMUM COELI (IC)** is located, is always in the same position in the wheel regardless of its Cusp Sign.

Left Of The Prime Meridian (Ascendant)

The left side of the prime meridian is home to the **ASCENDANT (ASC)**, or Rising Sign, and is located on the **EASTERN SECTOR** of the wheel, representing the initial spark that is the dawn of your life. As you become involved in work and other life activities, you present yourself as you want the world to see you, but not necessarily as you really are.

The Ascendant is always located at the first house, which is ruled by Aries, and is at 6AM on the astrological chart. The sign that falls in this house on the day and time that you were born drives your personality persona. Think of this location as being the first source of light that shines upon how you're seen by others.

HOUSE ONE, where the **ASCENDANT (ASC)** is located, is always in the same position in the wheel regardless of its Cusp Sign.

Right Of The Prime Meridian (Descendant)

The right side of the prime meridian is the location of the **DESCENDANT (DESC)**, which is receptive in nature and represents balance between your inner self and the outer world in relationships or the issue at hand, e.g., a marriage or even a legal partnership.

Located at 6PM in the seventh house of Libra, the Descendant reveals the type of close relationships you're drawn to. This location can be thought of as the point where as the Sun sets, it enhances your ability to express yourself more freely with an intimate partner in your life.

HOUSE SEVEN, where the **DESCENDANT (DESC)** is located, is always in the same position in the wheel regardless of its Cusp Sign.

Summary of Wheel Divisions, Direction, and Houses

EAST IS ON LEFT SIDE OF THE WHEEL. The first house cusp, or division, called the Ascendant (ASC), is the starting point of the day in astrology. It corresponds to the sunrise point. The first house in the chart shows how the world sees you. And internally it corresponds to your persona, i.e., how you present yourself to the world as a personality (persona). The ruler of the first house is Aries. Thus, regardless of the sign on this house, there is an underlying influence of the energy of Aries.

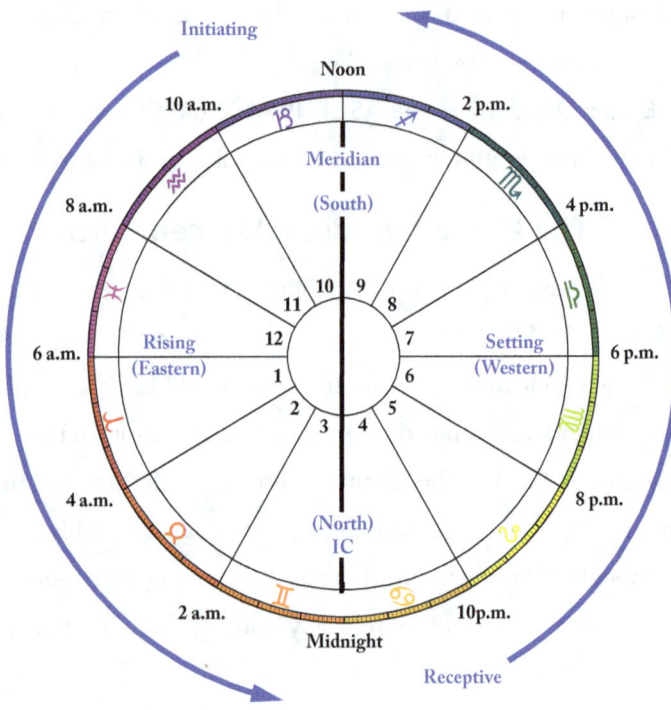

Illustration 44: Reviewing the Wheel

SOUTH IS AT THE TOP OF THE WHEEL. On the tenth house cusp, the top of the wheel, is the Medium Coeli (MC); this house represents your reputation in the world, your goals, and dreams, and how you show yourself to the public, i.e., your career and your way of being in a profession, your position of authority. The ruler of the tenth house is Saturn. Thus, regardless of the sign on this house, there is an underlying influence of the energy of Capricorn.

WEST IS ON RIGHT SIDE OF THE WHEEL. The seventh house cusp or the Descendant (DESC) is opposite the Ascendant and thus marks the beginning of evening in astrology. It reveals how you interact with those with whom you are closest like a marriage partner, and it denotes balance and love. Legal relationships show up in this house. The ruler of the seventh house is Venus. Thus, regardless of the sign on this house, there is an underlying influence of the energy of Libra.

NORTH IS AT THE BOTTOM OF THE WHEEL. The fourth house is the most internal house in the wheel. It marks the point of the Imum Coeli (IC) and shows the early psychological foundations to your core of being and shows something of your experience growing up and what influenced you, your vulnerabilities and sensitivities. The ruler of the fourth house is the Moon. Thus, regardless of the sign on this house, there is an underlying influence of the energy of Cancer.

The Importance Of Longitude & Latitude

Because on Earth, time measurement is based on the motion of the Earth around the Sun, it is important to know something about how time is measured to understand astrology charts.

88 ♦ ASTROLOGY ESSENTIALS

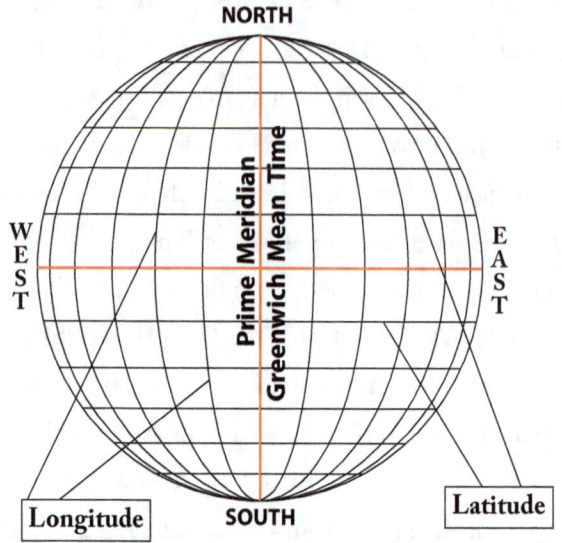

Illustration 45: Latitude and Longitude

The Earth is divided into two sets of imaginary lines called Longitude and Latitude. Longitude lines go around the globe from North to South vertically. They begin at the Prime Meridian or zero degrees (0°) longitude, which is arbitrarily set at Greenwich, England, known as the Prime Meridian, and referred to as Greenwich Meantime (GMT).

Longitude determines the exact time at a specific place on the globe and this is referred to as the Time Zone. This construct helps astrologers adjust a birth chart to the exact position of the planets in the chart based on the time at a particular location specifically.

Longitude Example

Because cities, towns, and other locations around the globe don't always land on a precise point of longitude, astrology must make adjustments (also known as *corrections*) in order to create charts that provide accurate information.

Using a rounded time zone (one that doesn't adjust for the exact location) does not give precise astrological positions of the planets in the wheel and therefore, the information provided for analysis in a chart will contain some inaccuracies. Computer programs automatically make this adjustment if you use the exact location for an event.

For example, the longitude of Los Angeles is: 118W15 (118 degrees west 15 minutes). To calculate exact time at the location of the astrological event a mathematical correction is applied to the Longitude position from GMT in order to adjust for the fact that Los Angeles isn't located exactly on a longitude Meridian. For example, when it's 12:00 PM in Greenwich, England, it's 4:00 AM in Los Angeles, but when we adjust for where LA actually rests on the planet, the time is actually 4:07 AM in that city.

To correct the time: 118°W15' is divided by 180° (half the globe) and is multiplied by 12 hours (half the time for the earth to rotate on its axis). GMT is the zero point reference. This division equals 7.883 which equals 7 hours 53 minutes when converted to actual time.

How Time Dramatically Affects a Chart
Astrological Chart Examples

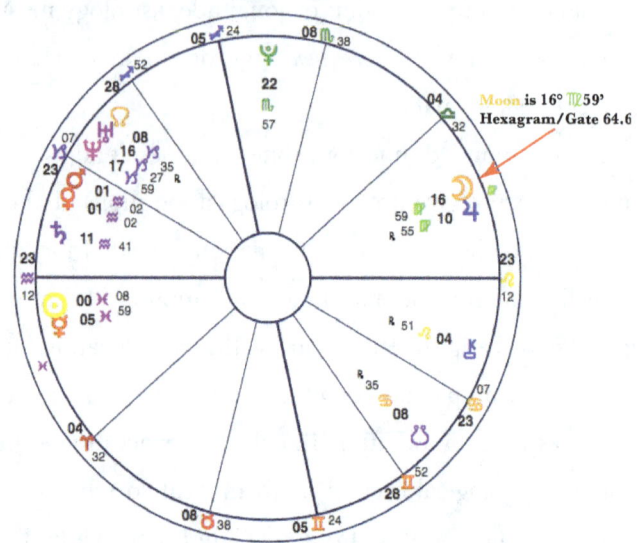

Illustration 46: AJS February 19, 1992, 6:53 am, San Antonio, Texas – How time affects an astrological chart

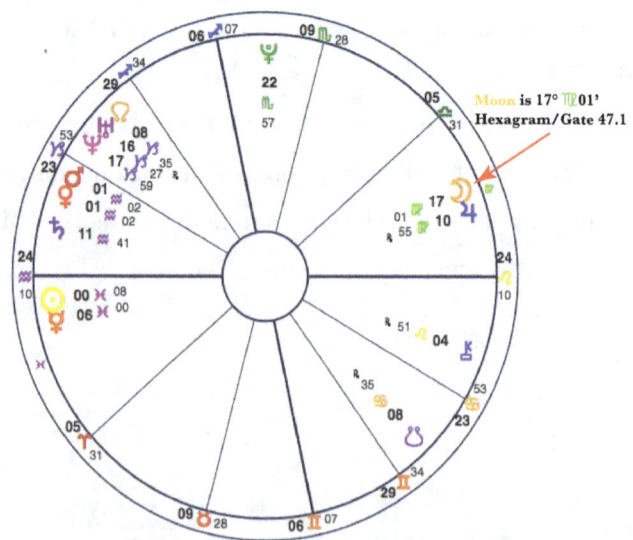

Illustration 47: AJS February 19, 1992, 6:56 am, San Antonio, Texas – How time affects an astrological chart

How Time Dramatically Affects a Chart: Noble Sciences Chart/Map Example

Example: A 2-Minute Time Difference changes the Moon position and the colors in the Chart/Map.

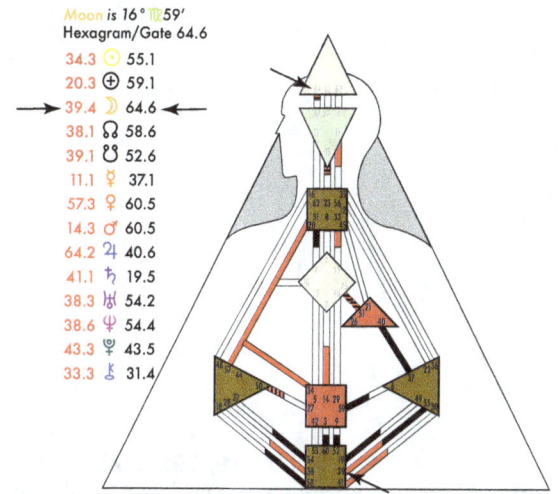

Illustration 48: AJS February 19, 1992. 6:53 AM, San Antonio, Texas – How time affects a Body Graph

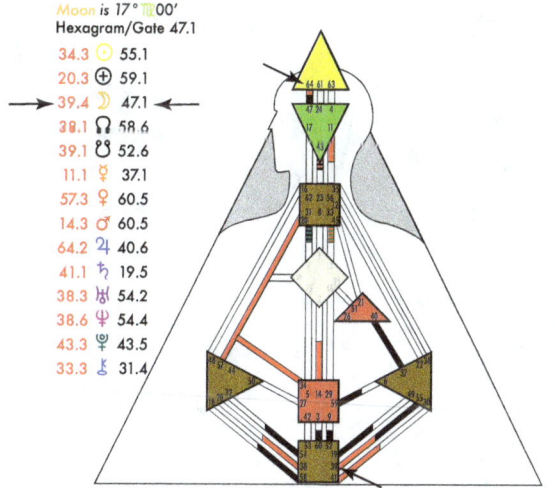

Illustration 49: AJS February 19, 1992, 6:56 AM, San Antonio, Texas – How time affects a Body Graph

92 ♦ ASTROLOGY ESSENTIALS

How Place Dramatically Affects a Chart: Astrological Chart Examples

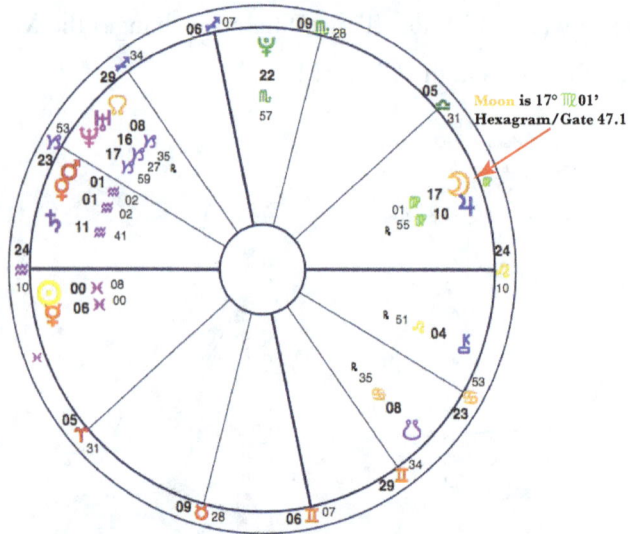

Illustration 50: AJS February 19, 1992, 6:56 AM CST, San Antonio, Texas – How place affects an Astrological Chart

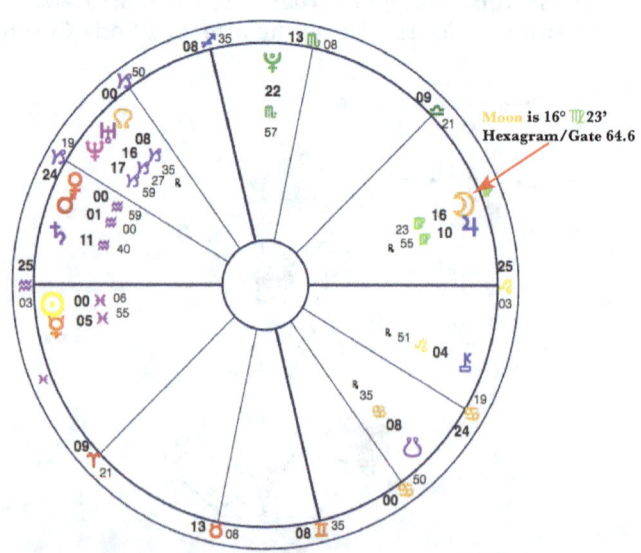

Illustration 51: AJS February 19, 1992, 6:56 AM EST, Charlotte, North Carolina – How place affects an Astrological Chart

How Place Dramatically Affects a Chart: Noble Sciences Chart/Map Example

Example: A 1-Time Zone Difference changes the Moon position and the colors in the Chart/Map.

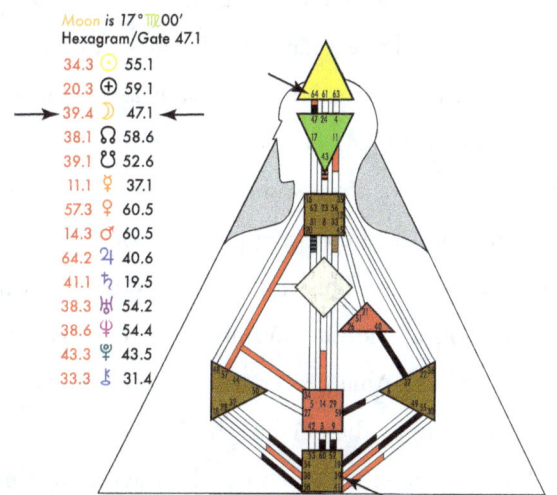

Illustration 52: AJS Body Energy Map – February 19, 1992, 6:56 AM CST, San Antonio, Texas – How place affects the Body Graph

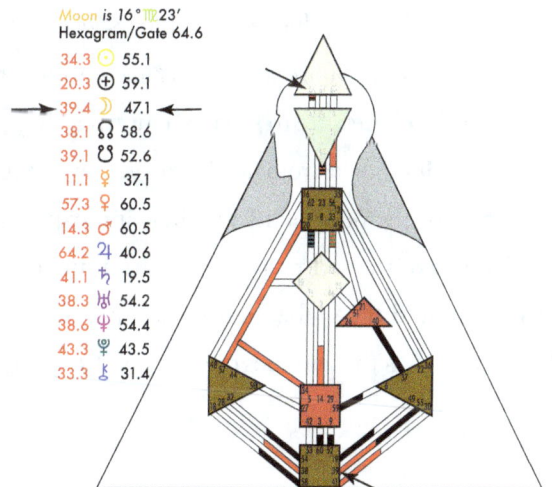

Illustration 53: AJS Body Energy Map – February 19, 1992, 6:56 AM EST, Charlotte, North Carolina – How place affects the Body Graph

Notice subtle shifts, especially in the Moon positions, in the example of how time and place in the astrology calculations changes a chart. Even small differences in time or place can significantly alter the Noble Sciences charts in terms of energy center definitions and thus, the precision of the data is essential in calculating an accurate chart. Small nuances in time and place have significance and must be taken into account for accuracy of interpretation in any astrological, Human Design, or Noble Sciences Charts.

In the charts shown in this example, the differences in time or place show a small shift in the degree positions of the moon in the charts and in the ascendant. This results in a difference in the body maps in the activation in the Crown and Ajna centers. The person with an open Crown and Ajna center without their activation is less likely to feel the pressure of the person who has both centers defined or activated. The person with the activated Crown and Ajna center is likely to feel pressure to make sense of things and to hold to their own perspective on things more than the person with the open centers. In addition, the person with the defined centers when connecting with energies that activate the Throat center is likely to function as a channel for others and feel pressure to speak about what they perceive. Thus, even a small difference astrologically can make a large difference in manifestation of the personality and how it is expressed.

The **LATITUDE** axis shows Imaginary lines running East to West **HORIZONTALLY** around the globe. Since the Equator is the widest part of the Earth, it's set as the zero degree (0°) point Latitude.

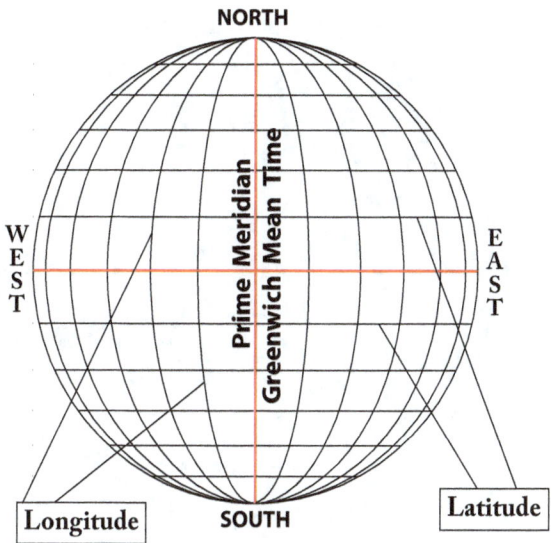

Illustration 54: Latitude and Longitude

Changes in Latitude shift planetary positions and house cusps. These kinds of shifts affect the rising sign, or Ascendant, changing where planets are in the astrological wheel.

Adjusting latitude also changes the degrees and minutes of some planetary positions, specifically, the area of life affected (the house). It also accounts for subtle planetary expressions in life and health, and changes the timing of events.

96 • ASTROLOGY ESSENTIALS

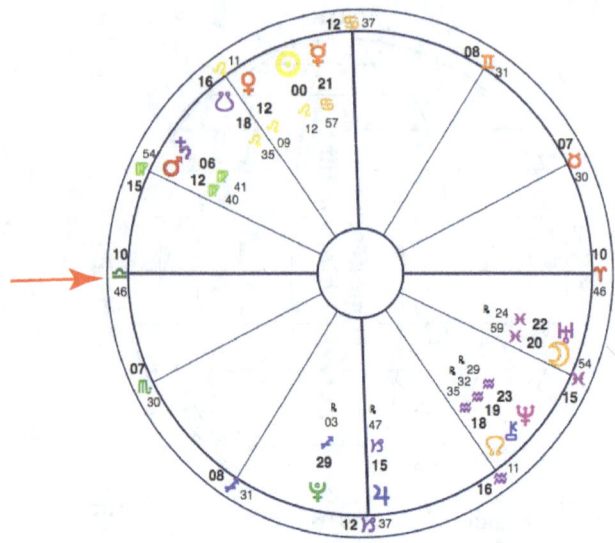

Illustration 55: Blossburg, Pennsylvania, July 22, 2008, 12:00 PM EDT, Latitude: 41°N40'44", Longitude: 077°W03' – How Latitude and Longitude affect an astrological chart

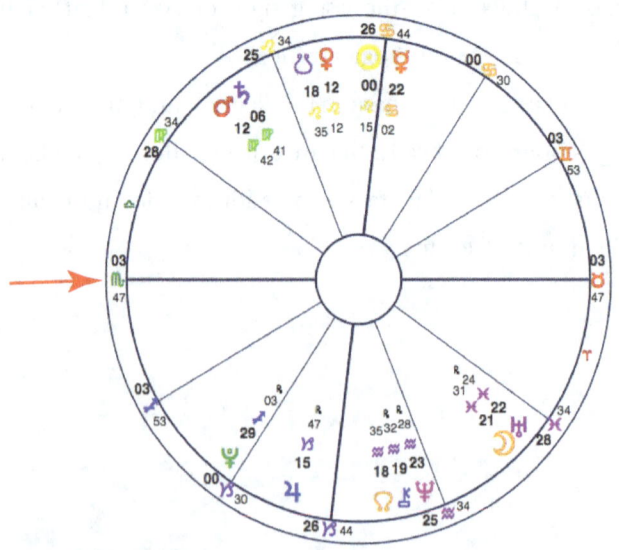

Illustration 56: Lima, Peru, July 22, 2008, 12:00 PM EDT, Latitude: 12°S03' Longitude: 077°W03' – How Latitude and Longitude affect an astrological chart

Someone with a 10 degree 46 minute Libra ascendant will likely be concerned with the aesthetics and balance in their life, and, in all probability will likely be attracted to and interested in the arts. They are likely to be diligent and active in their pursuit of the arts and will feel ill at ease when balance and harmony is absent in any situation or relationship. The 10-degree position of Libra connects in a square (an 88 degree aspect) with the signs and planets in the midheaven as well as with the Jupiter position in Capricorn (also a square) because of the degree of the ascendant. Notice that all three of the square aspects are in Cardinal signs.

Someone with 3 degrees 46 minutes of Scorpio on the Ascendant is likely to feel things intensely, be interested in consciousness, have a healthy sexual drive, and expect others to understand their depth. They will have a tendency, because of the actual degree, to be optimistic in their outlook and to feel strongly that their ideas and opinions matter and should be communicated to others.

When reading an astrological chart, the ruler of the sign is taken into account as well as the nuance of the degree of the sign. Libra in the first example chart is ruled by Venus and is happy in the second and seventh house of the chart. The second house relates to self-esteem and resources and the seventh house relates to relationships so these considerations go into a reading. Scorpio, in the second example, is ruled by Pluto, a planet that rules the eighth house. Based on advanced astrology, three degrees of Scorpio relates to Sagittarius, in nuance, and thus, this degree also relates to Jupiter, the ruler of Sagittarius. The planets in those signs are considered to have an influence on the personality and outlook shown in the chart.

The Body Maps do not consider the Ascendant in their graphics, thus, change is less visible than in the astrological chart maps. Nevertheless, it is important to recognize that small shifts in time and place of birth affect the nuanced components of a person's chemistry and psyche. In order to be accurate in all charts and to be able to work with accurate information it is essential to be precise and to calculate charts using exact information.

Basics of An Astrological Calculation

Astrology always works with a 360-degree wheel of the Zodiac. Each of the 12 astrological signs measures 30-degrees in the circle. Aries starts at 0-degrees, Taurus at 30-degrees, Gemini at 60-degrees, Cancer at 90-degrees, and so on. Pisces starts at 330-degrees.

In order to calculate an accurate astrological chart, you must use mathematics to convert every degree into its accurate zodiac degree, minute, and second. Although the computer calculates the charts for you, it is important in moving forward with Noble Sciences Calculations of Chart/Maps and Layers to understand a little of how the mathematics works. For example: If someone's birthday is March 15, 1938 at 2:54 PM in Detroit, Michigan, their Sun is positioned at 24° 34' 54" (24-degrees, 34-minutes, 54-seconds) in the Sign of Pisces.

If you look at the graphic on the next page, you can see that Pisces begins at 330°.

To Calculate the Zodiac Wheel Position of the Birth Date:

330° 00° 00° (Wheel Position)
+ 24° 34' 54" (Pisces Position)
354° 34' 54" (Sun's Position)

Noble Sciences calculates the Zodiac position of the Sun and of the Moon to find other positions used in Noble Sciences Charts/Maps. The Event under consideration is the Natal or Birth time in the case example. This time is used for the Natal Sun position of the Charts/Maps. In this case example, the Natal Sun is at: 24°34'54" of Pisces.

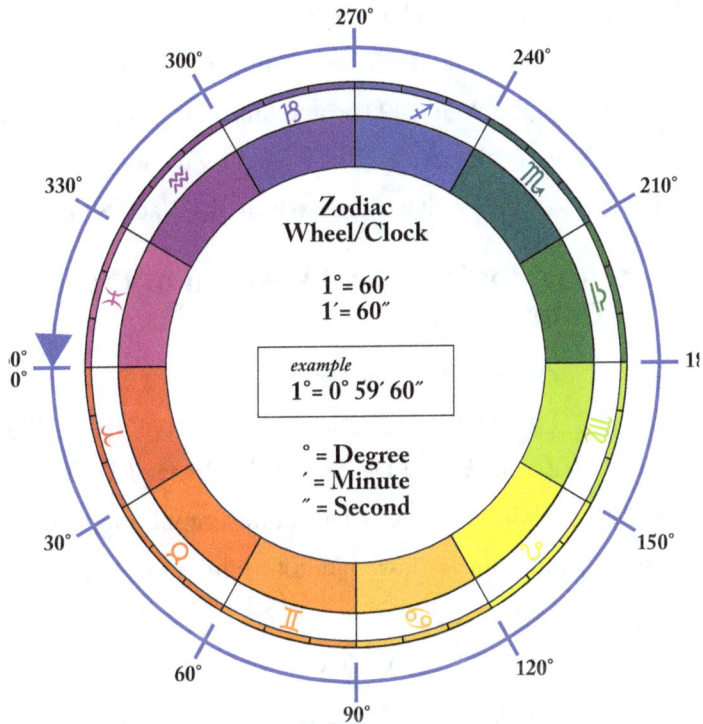

Illustration 57: The Astrological Signs converted to degrees in the Astrological Clock

As an example, consider the way Noble Sciences calculates the Prenatal Sun, (the Sun before birth). At birth the Sun was 24°34'54" (24-degrees thirty-four minutes, 54-seconds of Pisces).

Noble Sciences uses 88 Degrees before the Natal (birth) date to calculate the Prenatal date and time. Since 88 degrees can't be subtracted from a 30 degree sign, the Astrological Sign Degrees are converted into their actual Zodiac wheel degrees in order to figure out the prenatal date and its time. Thus, a birthday of March 15, 1938 at 2:54 PM in Detroit, Michigan puts the Sun at 24° 34' 54" (24-degrees, 34-minutes, 54-seconds) of Pisces.

Astrological Calculation Mathematics for Charts/Maps

The Sun is at 24° 34' 54" of Pisces or 354° 34' 54". Pisces begins at 330° 00' 00" , so you add the Sun's position 24° 34' 54".

To determine the Prenatal Date and time to use in the Noble Sciences Chart/Map for this person, calculate the Zodiac wheel position by subtracting 88° (eighty-eight degrees) from the Natal Sun position.

Calculate the Zodiac Wheel Position of the Prenatal Date's Sun:

354° 34' 54 " (Zodiac Wheel Degrees) of Natal Sun
  88° 00' 00 " (Prenatal degree position is 88° before the natal Sun)
266° 34' 54"   (Zodiac Wheel Degrees of the Prenatal Sun)

To figure out what sign this is, look at the graphic on the previous page to see the indicator for less than 266° – in this case 240°. Sagittarius begins at 240°.

266° 34' 54 " (Zodiac Wheel Degrees)
240° 00' 00 " (Where Sagittarius begins)
  26° 34' 54 " (Sagittarius Position)

The Prenatal Sun is at 26° 34' 54" of Sagittarius ♐. By looking at a table of planetary positions for the date and time calculated, the prenatal date is December 18, 1937 at 21:46:55 GMT.

Simple Chart Example

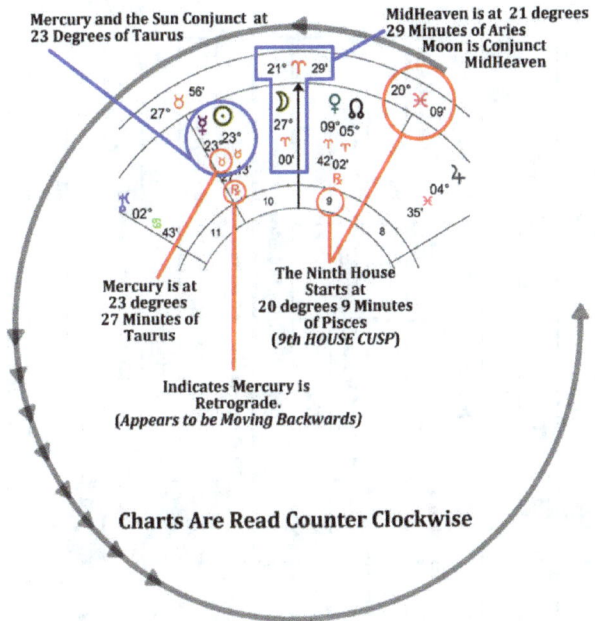

Illustration 58: An Astrological Chart Example

When two planets are close to each other by a few degrees they are said to be conjunct. This example shows Mercury conjunct the Sun at 23 degrees of Taurus in the tenth house. With Mercury and the Sun so close to each other we know that this individual likes to understand things and grasp them to feel a sense of accomplishment and security. Taurus is an earth sign in an earth house.

The Moon is on the Midheaven (tenth house cusp) in the sign of Aries. The Moon, thus, has an initiating active energy in a house related to how the individual expresses him/herself in the world and in profession. It also shows that on an emotional level, a sense of accomplishment is important to this person. The Moon is in a fire sign in an earth house giving it a bit of dynamic energy in its expression.

Key to Colors and Symbols

Sign Name and Glyph		Planet Name and Glyph		Center Color/Shape Center Name	Trigram	Keynote
Aries	♈	Mars	♂	Crown (Head)	☳	Thunder
Taurus	♉	Venus	♀	Ajna	☷	Earth
Gemini	♊	Mercury	☿	Throat	☶	Mountain
Cancer	♋	Moon	☽	Self	☵	Water
Leo	♌	Sun	☉	Heart	☴	Wind
Virgo	♍	Mercury	☿	Sacral	☰	Heaven
Libra	♎	Venus	♀	Splenic	☱	Lake
Scorpio	♏	Pluto	♇	Solar Plexus	☲	Fire
Sagittarius	♐	Jupiter	♃	Root		
Capricorn	♑	Saturn	♄	Body Graph Colors		
Aquarius	♒	Uranus	♅	Mental/Interactive Reality Earth World		
Pisces	♓	Neptune	♆	Spiritual/Archetypal Sleep/Dream World		
		Chiron	⚷	Emotional Mystical Inspiration/Angelic World		
		Earth	⊕	Physical/Hormonal Circadian Biological World		
		North Node	☊	Active Gate		
		South Node	☋	Receptive Gate		
				Chiron Activated Gate		

Types: Manifestor – Generator – Manifesting Generator – Projector – Reflector

Illustration 59: Color Key

ASTROLOGICAL CALCULATIONS ♦ 103

♈ Aries	♉ Taurus	♊ Gemini
♋ Cancer	♌ Leo	♍ Virgo
♎ Libra	♏ Scorpio	♐ Sagittarius
♑ Capricorn	♒ Aquarius	♓ Pisces

☉ Sun	☽ Moon	☿ Mercury
♀ Venus	♂ Mars	♃ Jupiter
♄ Saturn	♅ Uranus	♆ Neptune
♇ Pluto	☊ North Node	☋ South Node
⚷ Chiron	⚴ Pallas	⚵ Juno
⚳ Ceres	⚶ Vesta	AS Ascendant
MC Midheaven	⊗ Part of Fortune	⊕ Earth

☌ Conjunctions	☍ Oppositions	△ Trines
□ Squares	⚺ Semi-Sextiles	♃ Sextiles
⊼ Inconjuncts	∠ Semi-Squares	⚼ Sesquiquadrates
Q Quintiles	✶ Septiles	⋈ Noviles
Q Quindeciles	// Parallels	# Contra-Parallels

Illustration 60: Astrology Legend

Planet	Time to Orbit Sun	Natal Return Cycle	Time in a Gate	Time in a Line
Sun	365.25 days about 1° per day	Yearly About 1°/day	5.7 days	.95 days
Earth	365.25 days about 1° per day	Yearly About 1°/day	5.7 Days	.95 Days
Moon	27.3 days about 12°/day	Lunar return is monthly	.46 days	.08 day
Mercury	87.96 days (always within 28° of the ☉ up to 2°30'/day)	About 3 months	1.4 days	.23 day
Venus	224.68 days 1°15'/day	About 7.5 months	3.5 days	.59 day
Mars	686.95 days (1.89 years) 0°40'/day	About 23 months	10.78 days	1.8 days
Jupiter	11.87 years about 30°/year	11.87 years	67.69 days 2.25 months	11.3 days
Saturn	29.46 years about 1°/year	29.46 years	168 days 5.6 months .46 year	28.03 days .08 year
Uranus	84.05 years about 4°/year	84.05 years	479.34 days 15.98 months 1.31 years	79.92 days .22 year
Neptune	164.81 years about 1° to 2°/year	164.81 years	940 days 31.33 months 2.58 years	156.77 days .43 year
Pluto	248.54 years about 1°/year Elliptical orbit not equal in all signs	248.54 years	1417.45 days 47.25 Months 3.88 years	235.6 days .65 year
Chiron	50.7 years	50.7 years	289.15 days 9.64 months .79 days	48.19 days .13 years
North Node	18.6 years about 20°/year retrograde	About 18.6 years	106.08 days 3.54 months .29 years	17.68 days .05 year

(Mean planetary orbits from Astronomy: From the Earth to the Universe. 3rd Edition. Jay Pasachoff. Sunders, NY. 1987.)

Illustration 61: Table of Planetary Movement

ASTROLOGICAL CALCULATIONS ♦ 105

0° Zodiac Position of Astrological Sign	Zodiac Symbol	Planetary Ruler by Astrological Sign	Hexagram & Line Number at 0° of Sign	Hexagram Glyph
Aries	♈	♂	25.2	25
Taurus	♉	♀	3.4	3
Gemini	♊	☿	8.6	8
Cancer	♋	☽	15.2	15
Leo	♌	☉	56.4	56
Virgo	♍	☿	29.6	29
Libra	♎	♀	46.2	46
Scorpio	♏	♇	50.4	50
Sagittarius	♐	♃	14.6	14
Capricorn	♑	♄	10.2	10
Aquarius	♒	♅	60.4	60
Pisces	♓	♆	30.6	30

Illustration 62: Hexagram Correspondences to Zero Degree Points of Each Zodiac Sign

Legend: Astrological Sign is Color Coded by its Sign at 0° point; Sign is Color Coded; Planetary Ruler of Sign is Color Coded by Astrological Sign; Hexagram Gate and Line is Color Coded by Color of Center in Body Graph; Hexagram is Color Coded by Astrological Sign of Lower Trigram of Self Center

Noble Sciences Tools
Interpretive Foundations in Astrology

Noble Sciences Tools use traditional Tropical Placidus calculations.

I studied the following disciplines extensively. They form the theoretical foundation of Noble Sciences astrology interpretations. I am ever grateful to my extraordinary teachers.

Katherine de Jersey's teachings including: keynotes, progressions, diurnals, timing, the importance of degrees of the Zodiac, declinations, and the Marriage of Science and Intuition, her "secrets" to accuracy.

Pamela Crane's teachings including: interdisciplinary astrology, asteroids, colors, dwads, converse progressions, harmonics, declinations, and creative integrations.

Gloria Stein's teachings including: astrological math, the essentials of a strong foundation, progressions, returns, relocation, and interpretive dynamics of/in a chart.

Thyrza Escobar's teachings including: the Star Wheel, dwads, superdwads, timing, developmental cycles, planetary cycles, the natural wheel.

Magi Astrology teachings including: the importance and interpretation of Chiron.

Additional Studies including: Builder's of the Adytum, Seminars, extensive reading, Individual Studies, etc.

About the Author
Eleanor Haspel-Portner, Ph.D.

Eleanor was born in Brooklyn, New York on December 11, 1944 at 11:10 AM EST. She received her Ph.D. from The University of Chicago, Department on Comparative Human Development, in one of the first interdisciplinary departments in the United States. Eleanor uniquely integrates her background and training in the Social Sciences (psychology, biology, anthropology, sociology) with esoteric studies.

By applying multidimensional **NOBLE SCIENCES TOOLS™** that she developed and validated, Eleanor helps people transform their lives. Throughout her extensive career in private practice as a licensed Clinical Psychologist, Reiki Master/Teacher, and Transformational Life and Relationship Coach, Eleanor helped thousands of individuals, couples, and groups synthesize their life experiences in practical ways for living healthy, successful, and creative lives.

Eleanor strongly believes that each individual's core Self manifests fully when given support and encouragement. She also believes that many people simply need some directional help to feel empowered in their lives. Eleanor and her husband, Marvin, work together in documenting **NOBLE SCIENCES TOOLS™**. They met in India on August 13, 1978 and have been married since then. They live closely together with their dog, cats, and pet ducks. They very much enjoy their children and grandchildren.

Eleanor's first book, "Marriage in Trouble: A Time of Decision" was published in 1976 by Nelson-Hall. In recent years her writing and publishing efforts have been focused on developing Noble Sciences Materials and Tools and self-publishing her articles, books, and graphics.

You can contact Eleanor at: ehp@noblesciences.com or call: (310) 230-7787.

Acknowledgments

When I first began exploring astrology in 1971, I thought it was the most challenging subject I had ever studied. At the time, I was living in Chicago and already had my doctorate with a strong background in statistics and metaphysics. Despite my intention to study astrology seriously, I could not find a teacher so I gave up the study, but not my interest. Thus, I was thrilled when, in 1973, a friend of mine arranged for me to have a professional astrology reading with Katherine de Jersey. Her reading was so accurate and so inspiring that I pursued readings with her for my children, friends, and clients.

Katie and I became lifelong friends; her influence on my life and its direction was instrumental. And, because she was such a powerfully accurate astrologer, I was motivated to learn her "secrets". To do so, I studied astrology for many years with the world's foremost astrologers while continuing to pursue my professional interests.

In 1996, my path led me to study Human Design, and as I became more educated in this system it expanded my understanding of how astrology and other disciplines show different facets of an individual's development. I recognized that any serious student of Human Design was best served by being well versed in astrological concepts so they could accurately interpret what is coded into the body graphs used in Human Design. Furthermore, as a social scientist, I researched the Human Design System statistically and found it was an incomplete system and many aspects of it did not hold up to scientific and/or astrological research scrutiny. Thus, my work moved me away from the Human Design community and onto my own path.

At the same time, and before her death, Katherine de Jersey asked me to do a reading on her using my "system." What Katie said about the reading spoke volumes. She said, "all these years as an astrologer, I thought I could see such depth, but this system you developed goes way beyond what I could ever hope to see in a chart. Do you think I am too old to study it?"

With this endorsement from Katie, I continued studying and expanding my work while testing and validating it clinically on clients. None of this work would have been possible without the support and assistance of family, friends, colleagues, and clients.

While I am grateful to all those who have trusted my work and its powerful insights, I also would like to call out to several people without whom, this book could not have found life.

First, I am deeply grateful to my beloved husband and soul mate, Marvin. He has been a constant support and inspiration, patiently reading all the documents I produce and adding immeasurably to them with his ideas and comments. My children and their families have supported my endeavors while serving as my closest research subjects.

Furthermore, many individuals contributed ideas and feedback to this beginning astrology book. Cindy O'Connor Smith worked closely with me on all the graphics in their initial iterations and tirelessly supported many incarnations of my process while developing images that speak volumes. Charles Haspel worked magic on the computer patiently leading me through my learning process while helping me gain confident skill in technology. Without Erik Memmert's Neutrino's through Windows and Simply Your Self computer program my work would not have seen the light of day. His support means so much to me and to the collective consciousness this work touches and honors.

For help bringing this volume to life, I want to thank Brijit Reed, Jones Pinsker, and Troy Kendall for their help editing and upgrading material. Michelle White's aesthetic sense and design acumen brought the book to its final format honoring the vision and vibration of the contents. Michelle has been a godsend in her patience and presence as well as in her expertise.

In Loving Light,

Eleanor Haspel-Portner, Ph.D.
Pacific Palisades, California

www.ingramcontent.com/pod-product-compliance
Lightning Source LLC
Chambersburg PA
CBHW070115080526
44586CB00013B/1303